<u>THE</u> CHLORINATION / CHLORAMINATION
HANDBOOK

WATER

DISINFECTION

SERIES

Gerald F. Connell

American
Water Works
Association

THE CHLORINATION / CHLORAMINATION
HANDBOOK

WATER

DISINFECTION

SERIES

The Chlorination/Chloramination Handbook
Water Disinfection Series

American Water Works Association
6666 West Quincy Avenue
Denver, CO 80235

Editor: Bill Cobban
Production Editor: Alan Livingston
Cover and Book Design: Scott Nakauchi-Hawn

Library of Congress Cataloging-in-Publication Data
Connell, Gerald F., 1929-
    The chlorination/chloramination handbook / Gerald F. Connell.
    xvi, 171 p. 19×24 cm. -- (Water disinfection series ; v. 1)
    Includes bibliographical references and index.
    ISBN 0-89867-886-2 (hardcover : alk. paper)
    1. Water--Purification--Chlorination. 2. Water--Purification--Taste and odor control. 3. Ammonia. I. Title. II. Series.
TD462.C67 1997
628.1'662--dc21   96-37817
                            CIP

ISBN: 0-89867-886-2

Contents

List of Figures

List of Tables

Foreword

About 23 years ago the American Water Works Association (AWWA) developed a manual to fill a need for increased understanding of the chlorination process in drinking water treatment. The result was AWWA Manual M20, *Water Chlorination Principles and Practices*. The manual was used to assist in training water treatment personnel and as a text for chlorination workshops conducted by AWWA, starting with the initial course at Cherry Hill, N.J., in 1972. The course and the manual were used extensively for almost 20 years.

This new handbook, expanded to include chloramination, compiles the latest information on chlorination and chloramination in order to help operators and engineers in the design and operation of water treatment plants.

The Chlorination/Chloramination Handbook is aimed at both design and operating levels, and contains basic information that will help those concerned with chlorination and chloramination processes. It presents information regarding recent developments ranging from trihalomethanes (THMs) to the fire codes.

The direction, support, and encouragement provided by the AWWA Disinfection Systems Committee are deeply appreciated. Particularly helpful were Charles N. Haas, Wayne B. Huebner, Mark M. Bishop, and Zaid K.K. Chowdhury, all committee members. Thanks also to Andrea M. Dietrich for her review of this book. At my company, Capital Controls Company Inc., the assistance and understanding of Dean Hertert, Tom Gadomski, Mike Stout, John Kreibel, Greg Kreibel, Blair Jones, Dianne Phelan, and Craig Tourtellot were much appreciated. At Severn Trent

Water, the help of Mike Farrimond and George Keay, and at The Chlorine Institute, Gary Trojak, must not go unrecognized.

Gerald F. Connell

This book is dedicated to my wife, Claire, whose support, understanding, and encouragement made it possible.

History of Chlorination and Chloramination

The statement, "Chlorination is an art, not a science" is often heard when a discussion focuses on the use of chlorine, the primary disinfectant of potable water. It is not known who first uttered these words or whether this is a quotation or simply a paraphrase of a recognized remark that seems appropriate to the history and practices in the water industry. However we wish to consider it, the statement applies quite adequately to past practices of chlorination and ammoniation in water treatment. Certainly it applies to the initiation of the chlorination and chloramination processes in the early part of this century and to the development of many of the current practices in the treatment of water.

Regardless of the origin of the statement, the use of chlorine and ammonia in water treatment evolved from simple beginnings to become an important part of the treatment process. As chlorine and ammonia continued to be used, more was understood of the processes, the results, and the advantages and disadvantages of their use. Eventually a set of design criteria, and treatment principles and practices for use were developed or evolved. This evolution resulted from the combined efforts of the industry, consulting engineers, regulatory personnel, health officials, academia, and suppliers of the chemicals and feed equipment.

1

Water Treatment

The chlorination or chloramination process is but one of the steps in the water treatment process. Each process step in the water treatment plant contributes to the production of potable, disinfected water. All of the nonchlorine and nonammonia processes used provide a reduction of the pathogen level in the water to reach the goal of suitability for human consumption. The process of adding coagulant, followed by settling and filtration, removes significant numbers of bacteria. This fact does not go unrecognized in the scientific and regulatory communities that have had an impact on the development of US Environmental Protection Agency (USEPA) regulations. The Surface Water Treatment Rule (SWTR), developed by the USEPA to meet the requirements of the Safe Drinking Water Act (SDWA) and its amendments, focuses on two processes—filtration and disinfection—to help improve the quality of the finished water. Disinfection implies chlorination, chloramination, or the use of other disinfectants, but does not necessarily mean filtration. Also, filtration does not necessarily imply removal of bacteria and other organisms, although it performs that function quite adequately. The SWTR places the filtration process at an equal level to disinfection (chlorination and chloramination) for the purpose of producing a water free from harmful organisms.

The fact that each part of the water treatment process contributes to the production of a disinfected, potable water may frequently be overlooked. All too often, chlorination has been considered as both the solution to and the creator of all problems—from the quality of the finished water to its clarity, color, and taste. While chlorination and chloramination may have some part in the process in today's water plant operation, the interaction and interdependence of one process with another must be understood and considered. Therefore, one must always evaluate the chlorination process in perspective with the total plant operation. The locations of points of chlorine and ammonia addition, the quantities of chlorine and ammonia needed, the length of treatment, the frequency of treatment, the disinfectant residuals desired, and the regulatory requirements are only a few of the factors to be considered in treatment plant design and operation.

Disinfection

Recorded history cites a disinfection step, the boiling of water, as early as 500 B.C., although there is no indication that the benefits of disinfection of the water were either understood or appreciated. Subsequently, an understanding was developed of the need to disinfect drinking water and the benefits to be gained. Many disinfection

Table 1-1

Disinfection
techniques

Method	Example
Physical	Heat, storage
Light	Ultraviolet irradiation
Metals	Silver
pH	Acids, alkalis
Oxidants	Chlorine, chlorine dioxide, ozone, iodine, chloramine
Others	Surface active agents

techniques were identified and their performance evaluated at many of
the treatment plants. Some disinfection techniques are shown in Table 1-1.

Waterborne diseases (e.g., typhoid, dysentery, and cholera) occurred
with regularity in the water systems of the United States in the 1800s.
The incidence rates were high in the early 1900s (e.g., more than 25,000
typhoid deaths in 1900) but decreased rapidly with the onset of
chlorination (less than 20 deaths in 1960) (Laubusch 1964). The data
provide similar results for other organisms. With such a dramatic drop in
illnesses and fatalities following the onset of chlorination and
chloramination, the need for feeding these chemicals was well justified.
To date, chlorine has emerged as the disinfectant of choice primarily
because of its effectiveness, efficiency, economy of operation,
convenience, and the persistence of a chlorine residual. The unique
properties of chlorine also make its use an acceptable technique for
taste-and-odor control, algae and slime control, main sterilization, and
many other purposes. Although chlorine has come under intense
scrutiny because some of the by-products of the disinfection process may
be cancer-forming agents, chlorine is expected to remain in wide use.
Perhaps this is another step in the evolutionary process that continues to
make chlorination an "art and not a science."

Chlorination

The first use of chlorine in water treatment (apparently in England
in the 1880s) appears to be the application of one of the hypochlorites or
chloride of lime. Hypochlorites, sodium and calcium, are chlorine
compounds that produce the disinfecting and oxidizing agent
hypochlorous acid (HOCl) when added to water. This process is
discussed further in chapters 3, 4, and 5.

The interest in using chlorine compounds spurred some work in
generating chlorine on-site by the electrolytic process. The electrolytic
process generates chlorine as sodium hypochlorite from brine solutions

by applying a direct current to the solutions. For the most part, this effort was related to treating wastewater (Stanbridge 1977).

Initial uses of chlorine appear to be limited to intermittent or occasional treatment for one-time solutions; the value of continuous chlorination and the presence of a chlorine residual in the finished water were apparently not recognized. In England, the first recognized use of chlorine in water treatment was in 1897 in Maidstone. The first recognized use on the continent was in Middelkerke, Belgium, in 1902. Continuous use of the chlorination process was first practiced in Lincoln, England, in 1905, and on the continent in 1902. These practices used either bleaching powder, at Maidstone; "chloros," a hypochlorite solution, at Lincoln; or a ferric chloride and calcium hypochlorite mixture at Middelkerke (Society for Water Treatment and Examination 1952).

In the United States, a similar path was being followed with the first recorded use of chlorine in Louisville, Ky., in 1896 when studies of the filtration process confirmed benefits from chlorine treatment. In this experiment, the electrolytic process was used. The first continuous use in the United States, at Boonton, N.J., in 1908, used hypochlorite (White 1992).

It was not until 1909 that liquid chlorine, or chlorine gas, as it is commonly called, was commercially available in the United States. Liquid chlorine was first used in Niagara Falls, N.Y., in 1912. In England, liquid chlorine was suggested as a source for use as early as 1903, but apparently not used until 1917 at Rye Common (Race 1918).

Progress in the use of this form of chlorine was limited until the development of satisfactory equipment to feed, measure, and handle the corrosive mixture of moist chlorine. By the 1920s, chlorine gas was on its way to replacing other forms of chlorine, which suffered from poor stability and varying chlorine content. The stability, quality, and purity of gaseous chlorine made its usefulness more desirable than the hypochlorites (Harald 1934).

Chloramination

Ammonia's reactions with chlorine and the formation of the chloramines were first noted prior to 1920. In 1916 the Ottawa, Ont., water treatment plant used aqua ammonia to produce chloramines as a means of reducing the cost of treating water. In Denver, Colo., the Denver Union Water Company experimented with mixtures of ammonium hydroxide and hypochlorite as a method of producing chloramines for water disinfection and to prevent aftergrowths. This work was conducted in 1917 at Lake Marston (Engineering News Record 1977A and B). Subsequent experimentation by the British Army Medical Corps determined that the use of chloramines also produced an

4

unexpected improvement in the taste and odor of waters that contained phenols (caustic poisonous crystalline acidic compounds) (Adams 1925). Previously the addition of chlorine to raw waters containing phenolic compounds produced an undesirable medicinal taste in the water due to the chlorophenols formed. Thus, an additional reason to use the chloramination process was now recognized—taste-and-odor control.

The first recorded use of ammonia with chlorine to prevent objectionable tastes and odors from phenols in water was at Greenville, Tenn., in 1926 (McAmis 1927). The addition of ammonia was limited to the use of aqua ammonia or ammonium salt solutions. Initial attempts to feed compressed liquid ammonia or ammonia gas met with considerable difficulty, due to the precipitation of calcium or magnesium salts at the points of ammonia addition. The precipitation was a result of the high pH levels caused by the ammonia. These salts clogged the feeding equipment making continuous feeding difficult. At about the same time, during the evaluation of ammonia addition, it was discovered that the chloramine residual was more persistent and less aggressive than the chlorine residual. These two characteristics led to the use of the chloramination process as a method to provide a disinfectant residual in long distribution systems or transmission mains. Several US cities used this practice into the middle of the century.

The discovery of the "breakpoint phenomenon" (described in chapter 3, page 33) in the 1930s raised some questions regarding the value of adding ammonia to water. It was found that this treatment process could contribute other taste-and-odor problems if chloramines were not properly controlled. Many cities discontinued chloramination due to these potential taste-and-odor concerns. During World War II, the scarce availability of ammonia further curtailed the use of the chloramination process (Kirmeyer 1993). These factors prompted the use of a free chlorine residual.

It was further established that some of the natural ammonia content of some surface and groundwaters produced their own chlorinous tastes. The solution to this condition was to use the breakpoint process, which involves the addition of sufficient chlorine to destroy the chloramines formed and provide a water that contains only a free chlorine residual and no chloramines.

Discovery of Trihalomethanes

As both the chlorination and chloramination processes gained further acceptance, new discoveries in these processes were limited to refinements in the existing methods and chemistry. It wasn't until the 1970s that the next significant development came about. This was the work done by Rook in Holland, who identified that the reaction of chlorine with organic materials dissolved in water produced a class of compounds called trihalomethanes (THMs). The organics, usually humic and fulvic acids that originate in decaying vegetative growths, can be found in agricultural runoffs, aquifers, and natural vegetation. The THMs were identified as cancer-causing agents. Their discovery in drinking water caused a concern about chlorination. A new round of developments in the evolutionary process was begun. Research began on the nature of the reactions producing THMs, the concentrations considered unacceptable in drinking water, and methods to reduce or prevent their formation.

The great rush of scientific work produced sufficient data to lead regulatory bodies in Europe and North America to set restrictive levels of acceptable THM concentrations in finished water and to determine ways to minimize its formation in existing plants. Concentrations of THMs in the United States were limited to 100 parts per billion (ppb) or 100 micrograms per litre (μg/L). Other countries in Europe and North America set similar limits (in Canada, for example, the limit was 300 μg/L).

Many economically undeveloped countries, however, were vitally interested in eliminating waterborne diseases, such as cholera and typhoid, and continued with the chlorination process regardless of THM concerns. History had shown that a properly chlorinated water dramatically reduced the occurrence of these diseases in the countries with water systems appropriately treated. In Latin America, the elimination of the risk of death and illness from cholera, ranging in the thousands and hundred of thousands, respectively, was enough justification to use chlorine.

US Regulations

The United States passed the SDWA in 1974 and added amendments in 1986 that set further criteria for acceptable levels of THMs. The act also encouraged alternate treatment methods to achieve the now twin goals of disinfection and minimized THMs.

In the early 1990s, the USEPA initiated a process to reach a consensus among the many interested parties. The "Reg-Neg process" has evolved into a regulation that will require a utility serving more than 10,000 customers to achieve THM levels of 80 μg/L or lower by the late 1990s and a further reduction to a level of 40 μg/L by the early 2000s if data are not developed to show that the 80-μg/L level will be tolerable to the public. A similar rule regarding haloacetic acids, as well as rules for utilities serving fewer than 10,000 consumers, will be developed. The actual numbers and final details of the rule were still under development at the time of this writing.

These regulations will cause utilities to reduce the use of chlorine. The use of chlorine at the raw water intake would be replaced by ozone or chlorine dioxide, and the point of addition would be moved closer to the clearwell or in the distribution system. Intermediate points in the treatment plant would be considered on an as needed basis. The addition of chlorine in conjunction with ammonia in the distribution system appears to have support.

European Regulations

Europe is undergoing a similar evaluation of the disinfection process and the role chlorine and ammonia will play in the treatment of water. While in the past, each country's regulations have set the requirement for water treatment, use of chlorine or other oxidants, the need for a residual, etc., the evolution of the European Community and its common regulations will require each country to meet the same standards. The Netherlands has water systems operating without a residual in the distribution systems, indicating its great concern about the formation of THMs and other chlorinated by-products. These compounds are also referred to as disinfection by-products (DBPs). Disinfection by-products are those compounds formed by the reactions of the disinfectant (e.g., chlorine, ozone, and chlorine dioxide) with organics in the water. Disinfection by-products also result from the decomposition products of the disinfectant (e.g., chlorite and chlorate).

The Netherlands' practices are not shared by the other European members. However, most of the common European practices in the treatment of surface waters involve pretreatment with either ozone or chlorine dioxide, followed by chlorination of the finished waters.

The Future

Chlorine and ammonia and chlorination/chloramination practices will continue to be under scrutiny as more is learned about the effects of the disinfection process and the resulting DBPs. Operation of surface water treatment plants and the quality of the water they produce will continue to be followed closely. Treated, potable water will continue to be examined in ever-increasing detail. New processes that will remove precursors will be discovered and used successfully. The more promising of these appears to be the use of granular activated carbon (GAC) or membrane filtration. Both processes enhance the use of chlorination, since the precursors would have been removed and the possibility of DBP formation would be minimized.

Whatever the development, it is expected that chlorination will survive and perhaps become enhanced as the process is more thoroughly understood and used more efficiently.

The entire water industry continues to work diligently to provide the quality water that the public deserves and demands. Water utilities have come to understand that their business, like any other business, is market driven. If the industry does not respond, its business will suffer. Industry executives recognize that alternatives to their product, such as bottled water and point-of-use or point-of-entry devices, will be fighting for their customers. The consumer is willing to pay more for high-quality water, and if the water utility does not provide it someone else will. Public confidence in potable water is paramount, as is the issue of public health.

As the 21st century approaches, the challenges to the industry will remain. Providing microbiologically safe water will remain vitally important. This is attainable by using chlorine and chloramines. One can speculate that the DBPs will remain a concern but of a secondary nature unless additional scientific information is developed. It is probable that the production of microbiologically safe water, as well as water free of DBPs, can best be achieved by the use of improved filtration and membranes to remove precursors.

References

Adams, B.A. 1925. *The Chloramine Treatment of Pure Waters*. Medical Officer.

Baylis, J.R. 1935. *Elimination of Taste and Odor in Water*.

Engineering News-Record 1977A. *Chloramine at Denver Solves Aftergrowth Problem*. Engineering News-Record (Aug. 2, 1977).

————. 1977B. *Permanent Plants for Chloramine Installed at Denver*. Engineering News-Record (Sept. 25, 1977).

Harald, C.H.H. 1934. *29th Annual Report*. Metropolitan Water Board, London.

Kirmeyer, G.J., G.W. Foust, G.L. Pierson, and J.J. Simmler. 1993. *Optimizing Chloramine Treatment*. Denver, Colo.: American Water Works Association Research Foundation and American Water Works Association.

Laubusch, E.J. 1964. Chlorination and Other Disinfection Processes. Washington, D.C.: The Chlorine Institute.

McAmis, J.W. 1927. Prevention of Phenol Taste with Ammonia. *Jour. AWWA*, 17(3):341–350.

Proceedings of the Society for Water Treatment and Examination. 1955. Vol. 4, pp 69–82.

Race, J. 1918. *Chlorination of Water*.

Stanbridge, H.H. 1977. *History of Sewage Treatment In Britain, Part 3, Chemical Treatment*.

White, G.C. 1992. *Handbook of Chlorination*, 3rd ed.

Properties of Chlorine and Ammonia

This chapter provides details of the physical and chemical properties of chlorine and ammonia gases and sodium hypochlorite, aqueous ammonia, and ammonium salt solutions. Safety requirements involving storage, handling, transportation, human physiological effects, and the recent influence of the fire codes on the use of these chemicals are covered in chapter 6.

Chlorine Gas

Chlorine gas is often referred to as *elemental chlorine*. Elemental chlorine does not exist naturally; it must be produced for use. By far the most common process for its production is the electrolysis of brine. Chlorine is produced, collected, purified, compressed, and cooled for storage and use. Chlorine is stored and shipped as a liquefied gas under pressure.

The purified, commercially supplied chlorine is dry (less than 150 parts per million [ppm] moisture content). The Chlorine Institute is currently redefining the term *dry chlorine* and the methods for analysis. The new definition will most likely be a sliding scale based on the temperature and phase of the chlorine (The Chlorine Institute 1995).

The chemical symbol for chlorine is Cl, which identifies one atom of the chemical. However, elemental chlorine exists as a two-atom molecule with the symbol Cl_2. The atomic weight is 35.46; the molecular weight is 70.92. Chlorine is neither explosive nor flammable, although it will support combustion and react violently with many substances. Chlorine exhibits a characteristic pungent odor that is irritating to the mucous membranes.

Liquid chlorine is amber in color and about one and one half times heavier than water at 88.8 lb/ft³ (1,422 kg/m³) at 60°F (15.6°C). It boils at −30.1°F (−34.5°C) at one atmosphere (atm). Gaseous chlorine is greenish yellow and about two and one half times heavier than air, or 0.200 lb/ft³ (3.21 kg/m³) at 34°F (1.1°C). The vapor pressure of chlorine versus temperature is shown in Figure 2-1. The curve illustrates the equilibrium existing between the gaseous and liquid states under pressure at varying temperatures. From this curve, the physical state of chlorine can be determined by knowing the pressure and temperature. If an artificial pressure condition is imposed, such as padding the container with dry air or nitrogen, the relationship to the curve would be influenced in a manner similar to the operation of a vaporizer (discussed in chapter 7).

Chlorine is only slightly soluble in water at 6.93 lb/100 gal (8.3 kg/m³) at 60°F (15.6°C). Solubility data are illustrated in Figure 2-2.

Chlorine is highly reactive. This high degree of chemical reactivity enables chlorine to combine with many other chemicals quite easily. As such, chlorine is recognized as a "building block" chemical, one of the world's basic chemicals. In the presence of moisture, chlorine is extremely corrosive and reacts rapidly with all common metals, such as iron and copper. Wet chlorine is unreactive with silver, tantalum, and some of the alloys of tantalum. In the dry state, chlorine is highly reactive with titanium, while very wet chlorine is compatible with titanium. Some plastics, such as Teflon® and Kynar®, can be used with gaseous chlorine whether wet or dry. Liquid chlorine is highly reactive with polyvinyl chloride (PVC), causing it to soften rapidly and affecting PVC's structural integrity. However, Teflon® and Kynar® can be used in liquid chlorine service. Gaseous chlorine, both wet or dry, will not react with PVC, chlorinated polyvinyl chloride (CPVC), or acrylonitrile butadiene styrene (ABS).

Table 2-1 summarizes some of the important physical properties of chlorine.

Chlorine gas is considered a respiratory irritant and is classified by the US Department of Transportation (DOT) as a poisonous gas. Concentrations above 0.3 ppm (by volume) in air can be detected by most people. In 1989 the Occupational Safety and Health Administration (OSHA) permissible exposure level (PEL) was set at 0.5 ppm, while the short-term exposure level (STEL) was set at 1.0 ppm. Court action overturned these levels. The only current OSHA regulation is the 1 ppm, time-weighted average (TWA) ceiling level. The National Institute of Occupational Safety and Health (NIOSH) issued the immediately dangerous to life or health (IDLH) value, which is currently 10 ppm.

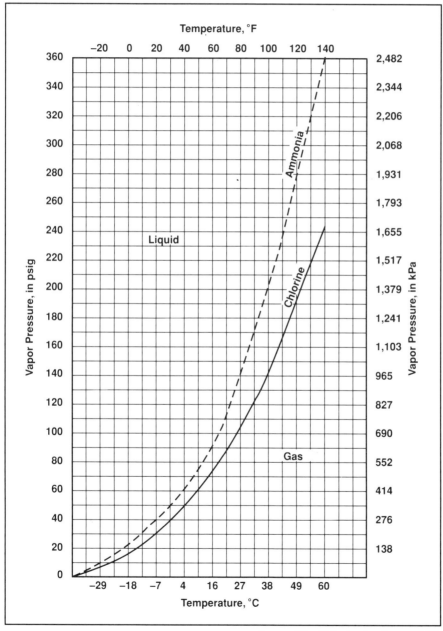

Figure 2-1

Vapor pressure of liquid chlorine

Source: The Chlorine Institute (1996).

Figure 2-2
Chlorine
solubility in
water

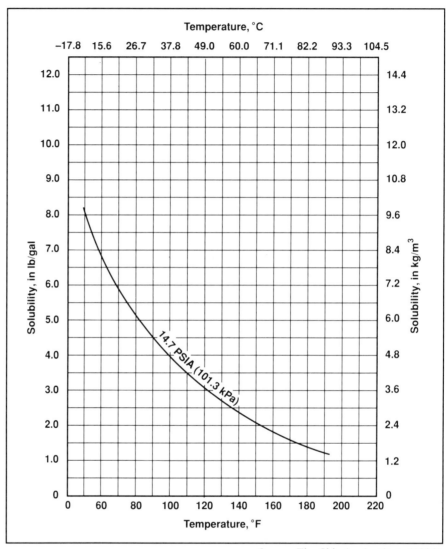

Source: The Chlorine Institute (1996).

Ammonia Gas

Ammonia is a gas at atmospheric pressure and ambient temperatures. Ammonia is not an element but rather a compound composed of the elements nitrogen and hydrogen. The ammonia molecule contains three atoms of hydrogen and one atom of nitrogen, has the chemical formula NH_3, and a molecular weight of 17. Small quantities of ammonia exist naturally. For commercial purposes,

Molecular weight	71.1 ·	**Table 2-1**
Boiling point at 1 atm (101.325 kPa)	−29.15°F (−33.97°C)	Physical properties of chlorine
Freezing point	−149.76°F (−100.98°C)	
Density (gas) at 34°F (1°C), 1 atm (101.325 kPa)	0.2006 lb/ft^3 (3.216 kg/m^3)	
Density (liquid) at 32°F (0°C)	91.56 lb/ft^3 (1.467 kg/m^3)	
Specific gravity (gas) at 32°F (0°C) 1 atm (101.32 kPa)	2.485 (air = 1)	
Specific gravity (liquid) at 32°F (0°C)	1.467 (water = 1)	
Solubility in water at 60°F (15.6°C)	6.93 lb/100 gal (8.30 kg/m^3)	
Specific heat (liquid) at 32°F (0°C)	0.2264 Btu/lb/°F (0.948 kj/kg/Kelvin)	
Heat of vaporization at −29.15°F (−33.97°C)	123.9 Btu/lb (288.1 kj/kg)	

Source: The Chlorine Institute (1996).

ammonia can be synthesized by a number of different processes that use hydrogen and nitrogen.

Ammonia has a pungent odor and, unlike chlorine, is colorless in both gaseous and liquid states. Ammonia liquefies at about the same temperature as chlorine but has a higher vapor pressure than chlorine at corresponding temperatures. The vapor pressure–temperature curve for ammonia is depicted in Figure 2-3. To illustrate the difference in vapor pressure between chlorine and ammonia with changing temperatures, the dotted line in the figure represents the chlorine vapor pressure curve. The vapor pressure of ammonia is about 20 percent greater than chlorine at ambient temperatures and increases at a more rapid pace than chlorine at higher temperatures. Unlike chlorine, reliquefaction of ammonia gas is not considered to be an operational problem. The boiling point or vaporization point of ammonia at atmospheric pressure is −28°F (−33.3°C).

Liquid ammonia is lighter than water, having a density of 42.57 lb/ft^3 (681.9 kg/m^3) at −18°F (−28°C). Ammonia gas is lighter than air with a density of 0.555 lb/ft^3 (0.8899 kg/m^3) at −18°F (−28°C). Ammonia is about 36 times more soluble in water than chlorine, 250 lb/100 gal (298 g/L) or 34 percent by weight at 60°F (20°C).

Ammonia is classified by the DOT as a corrosive gas. Chemically, ammonia is a relatively stable chemical compound and reactive only under specific conditions or with certain chemicals. Most common metals are unaffected by dry ammonia. In the presence of moisture or when dissolved in water, ammonia will attack copper and zinc and alloys containing these metals (e.g., brass and bronze). Ammonia will react with organics and inorganics to form ammonium salts. Reactions with chlorine can produce dangerous and/or explosive compounds, such as nitrogen trichloride. Physical data are illustrated in Table 2-2.

Figure 2-3

Vapor pressure of ammonia

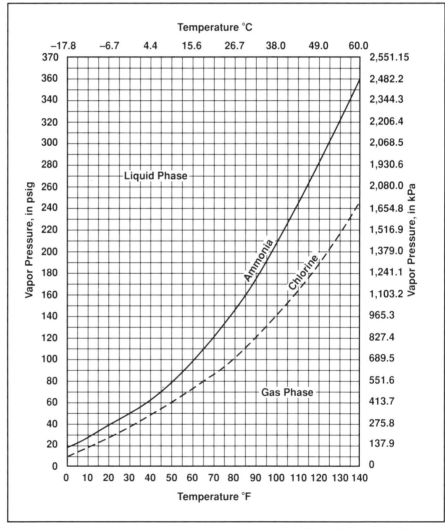

Source: National Ammonia Company.

Molecular weight	17.03	**Table 2-2**
Boiling point	$-28.2°F$ ($-33.4°C$) at 1 atm	Physical
Freezing point	$-107.9°F$ ($-77.7°C$)	properties of
Density (gas)	0.0555 lb/ft^3 (0.8899 kg/m^3) at $-27.7°$ ($-33.2°C$)	ammonia
Density (liquid)	42.57 lb/ft^3 (681.9 kg/m^3) at $-27.7°$ ($-33.2°C$)	
Flammability	Only within a range of 16 percent to 25 percent	

Source: Compressed Gas Association (1984).

Ammonia is considered a respiratory irritant, although it does not have a cumulative effect. Ammonia's pungent odor provides an early warning to alert the individual to its presence. Although individual physiological responses may vary, the least perceptible odor is considered to be 5 ppm (by volume). The current (1993) OSHA-PEL level is 35 ppm, while the American Council of Government Industrial Hygienists (ACGIH) threshold limit value and STEL values are 25 and 35 ppm, respectively.

Sodium Hypochlorite

Sodium hypochlorite is produced by reacting chlorine and sodium hydroxide as follows:

$$Cl_2 + NaOH \rightarrow NaOCl + NaCl + heat \qquad (2\text{-}1)$$

Sodium hypochlorite can also react with water to produce hypochlorous acid (Eq 2-2) just as chlorine gas produces hypochlorous acid by the hydrolysis of chlorine in water (refer to Eq 3-1 on page 24). As we will see, the active disinfecting or oxidizing agents, hypochlorous acid or hypochlorite ion, are available from either sodium hypochlorite or chlorine gas.

$$NaOCl + H_2O \rightarrow HOCl + NaOH \qquad (2\text{-}2)$$

The reaction indicated in Eq 2-1 proceeds by the addition of chlorine gas to a water solution of sodium hydroxide. Sodium hypochlorite produced is normally produced as a solution.

Sodium hypochlorite solutions are sometimes referred to as liquid bleach or javelle water. Some properties of different strength hypochlorite solutions are identified in Table 2-3. Generally the commercial or industrial-grade solutions produced have hypochlorite strengths in the 10 to 15 percent range. Sodium hypochlorite solutions are defined by many different terms, as defined by the following equations:

$$\text{trade percent} = \text{grams/litres available chlorine}/10 \qquad (2\text{-}3)$$

$$\text{weight percent} = \text{trade percent/specific gravity of the solution} \quad (2\text{-}4)$$

$$\text{available chlorine} = \text{pounds/gal} \times 0.083 \qquad (2\text{-}5)$$

Table 2-4 provides some of these data for a number of the more commonly used hypochlorite solutions.

Above 15 percent, the stability of hypochlorite solutions is poor, and decomposition and the concurrent formation of chlorate is of concern. Thus, higher concentrations are not normally available. Low concentrations of 5.25 percent or less are supplied as common household bleach (Bommaraju 1994).

The stability of sodium hypochlorite solutions is a function of the initial hypochlorite concentration at storage, temperature of storage,

Table 2-3
Hypochlorite solution strengths

Trade Percent	Density		Available Chlorine	
	lb/gal	*kg/L*	*lb/gal*	*g/L*
1	8.45	1.014	0.083	9.956
5	8.92	1.070	0.42	50.38
10	9.50	1.139	0.83	99.56
12	9.74	1.168	0.99	118.75
13	9.86	1.183	1.08	129.54
15	10.1	1.211	1.25	149.93

Source: The Chlorine Institute (1990).

Table 2-4
Physical properties of sodium hypochlorite

	Weight, *percent*				
	5	7.5	10	12.5	15
Specific gravity[*]	1.076	1.109	1.142	1.175	1.206
Density (lb/ft^3)[*]	67.14	69.20	71.26	73.32	75.25
(kg/L)[*]	1.076	1.109	1.142	1.175	1.206
Freezing point ($^{\circ}$F)	28.8	25.5	23.3	21.0	20.9
($^{\circ}$C)	−1.5	−3.6	−4.8	−6.1	−6.2

Source: The Chlorine Institute (1996).

[*]At 68°F (20°C).

length of storage, impurities present in the finished product, and exposure to light. The instability of the hypochlorite solution must be recognized when determining the feed rate and dosage.

Hypochlorite solutions do have a vapor pressure, which means that there is a gas released from the solutions (Table 2-5). The gas generally has been considered to be chlorine because the chlorine odor is readily detectable in hypochlorite storage and use areas. However, recent studies identified the off-gas as chlorine monoxide. Whatever the off-gas is called, in the presence of moisture the vapors will cause corrosion.

Of more importance to the water industry recently is the recognition of the chlorate produced in the manufacturing process and during storage. Sodium hypochlorite decomposition starts from the moment of production. The decomposition rate is a function of the previously mentioned parameters. The product of the decomposition is chlorate.

Recent studies indicate that chlorate and chlorite can cause hemolytic anemia in the blood. Toxicological studies are currently under way to determine the health significance of chlorate in blood. There will be some restriction on the amount of chlorate in the finished water that could affect the use of sodium hypochlorite. Since chlorate and chlorite are present in chlorine dioxide, there is a similar concern for these materials produced in chlorine dioxide reactions, decomposition, and generation.

The rate of decomposition of sodium hypochlorite has been studied by many researchers. The amount of decomposition product 'chlorate' formed is a function of the factors affecting stability. The amount of chlorate formed in the decomposition is shown in Figures 2-4 and 2-5. The percent chlorate formed is plotted as a function of time versus storage temperature in Figure 2-4 and as a function of initial hypochlorite strength in Figure 2-5. The amount of chlorate can be significant as the curves indicate levels up to 2 percent or more after 40 days. Its impact in regard to the finished water can be significant. For example, a hypochlorite solution containing 1,000 mg/L (ppm) (or 0.1

Temperature		Vapor Pressure		
°F	°C	mm Hg	psia	
48.2	9	3.7	0.071	
60.8	16	8.0	0.15	
68.0	20	12.1	0.23	
89.6	32	31.1	0.60	
118.4	48	100.0	1.93	

Table 2-5

Vapor pressure of hypochlorite solution[*]

[*]Solution strength: 12.5 percent by weight.

percent) of chlorate will add 1 mg/L (ppm) of chlorate to the treated water for each part per million of hypochlorite solution added. Permissible levels of chlorate have not been established as yet pending toxicological studies.

Chlorate formation can be defined by the following equations:

$$3NaOCl \rightarrow 2\ NaCl + NaClO_3 \qquad (2\text{-}6)$$

$$NaOCl + NaClO_2 \rightarrow NaClO_3 + NaCl \qquad (2\text{-}7)$$

Hypochlorite solutions will also react in a violent fashion with acids evolving chlorine gas. The reaction depends on the amounts of the chemicals mixed together. Ammonia can also react with hypochlorite solutions to produce chloramines, such as nitrogen trichloride, an explosive chemical at these concentrations. Organics, such as greases, oils, and fuels, can also react violently with hypochlorite, particularly after the product has dried and remains as a salt. Some physical properties for hypochlorite solutions are presented in Tables 2-3, 2-4, and 2-5.

There are other hypochlorite forms that have been used in water treatment, including calcium hypochlorite and lithium hypochlorite. Only calcium hypochlorite is of significance in drinking water. Currently its use appears to be limited to small systems. The reaction of calcium

Figure 2-4
Effect of temperature on chlorate formation in 13.1 percent hypochlorite

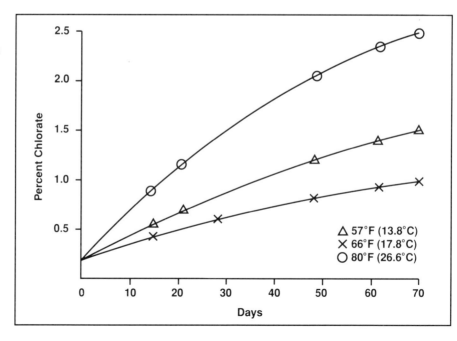

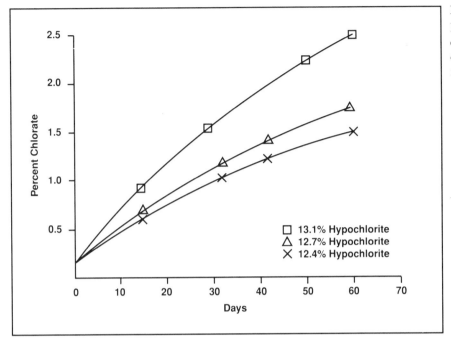

Figure 2-5
Effect of concentration on chlorate formation

hypochlorite in water is shown in the following equation:

$$Ca\,(OCl)_2 + 2\,H_2O \rightarrow 2\,HOCl + Ca(OH)_2 \qquad (2\text{-}8)$$

Because of the limited use of calcium hypochlorite, an interested reader should refer to publications produced by manufacturers.

Ammonia Solutions

Ammonia is also available in forms other than a compressed gas in cylinders. The most commonly used form is a solution of ammonia dissolved in water, usually referred to as aqueous or aqua ammonia. Other forms of ammonia are salts of ammonia solutions. Of these, the most frequently used is ammonium sulfate. The reaction of ammonia with water is shown in Eq 2-9.

$$NH_3 + H_2O \leftrightarrow NH_4OH \qquad (2\text{-}9)$$

The product of this reaction is ammonium hydroxide. Ammonium hydroxide is also produced from solutions of ammonia salts, as shown in Eq 2-10.

$$(NH_4)_2SO_4 + 2H_2O \rightarrow 2NH_4OH + H_2SO_4 \qquad (2\text{-}10)$$

Table 2-6	Specific gravity at 60°F (15°C)	0.8957
Properties of aqueous (aqua) ammonia, 30 percent by weight	Density at 60°F (15°C)	26.31°Bé (Baume) 55.7 lb/ft^3 (0.893 kg/L)
	Boiling point at 1 atm (101.325 kPa)	82°F (27.8°C)

Source: LaRoche Industries.

Eq 2-9 is reversible and will release ammonia as a gas. In addition, solutions that produce ammonium hydroxide will also provide ammonia. Since ammonia is available from these solutions, ammonium salt solutions are used as a source of ammonia just as sodium hypochlorite is used as a source of chlorine. All ammonium salt or aqueous ammonia solutions have alkaline pHs. The exact pH is a function of the solution concentration and temperature.

Since there is some ammonia gas evolved, a vapor pressure exists in aqueous ammonia and ammonium salt solutions, and all storage facilities must be suitably vented. Table 2-6 contains some information about the properties of aqueous ammonia.

Both ammonium salt solutions and aqueous ammonia exhibit the same characteristics that would be expected from ammonia. The solutions are to be treated as a source of a respiratory irritant and handled with care. Skin irritation and redness will develop if the solutions are spilled on the skin.

For additional information on the properties and handling of chlorine, ammonia, and their solutions, see chapter 6. Also ask the chemical supplier to provide the latest available material safety data sheet (MSDS) for the appropriate chemical.

References

Bommaraju, T.V. 1994. *Sodium Hypochlorite: Its Chemistry and Solubility*. Chlorine Institute Packaging Seminar, Minneapolis, Minn., February 1994.

Chlorine Institute. 1995. *Pamphlet 100, Sodium Hypochlorite Handling and Storage Guidelines*. Washington D.C.: The Chlorine Institute.

———. 1990. *Pamphlet 90, Toxicity Summary for Chlorine and Hypochlorites and Chlorine in Drinking Water*. Washington, D.C.: The Chlorine Institute.

———. 1996. *Chlorine Manual*. 6th ed. Washington, D.C.: The Chlorine Institute.

Compressed Gas Association. 1984. *Anhydrous Ammonia, Pamphlet G-2*. Alexandria, Va.: Compressed Gas Association.

LaRoche Industries. 1989. *Ammonia Technical Data Manual*. Atlanta, Ga.: LaRoche
    Industries.

Storage and Handling of Anhydrous Ammonia, Technical Data, February 1989.
    National Ammonia Company.

Chemistry of Chlorine, Ammonia, and Their Compounds

The treatment of water with chlorine and ammonia involves basic principles of chemistry that will be outlined in some detail in this chapter. The use of the words *chlorine* and *ammonia* in the following discussions and throughout this book relates to either substance in compressed gas form or the appropriate chlorine or ammonia solutions.

Chlorine and Water

There are three reactions in which chlorine participates in water treatment. These are oxidation, substitution, and disinfection. Oxidation is the reaction between two chemicals in which an exchange of electrons between one chemical and the chemical being oxidized takes place. Specifically, the oxidant, chlorine, gains electrons from the chemical to be oxidized. Substitution is the replacement of an element or portion of a chemical molecule by another ion, in this case, chlorine. Disinfection is the destruction of undesirable living organisms in the water.

When chlorine is added to water a chemical reaction occurs between the molecules of chlorine and water. Chlorine reacts rapidly with water to form two separate and distinct chemicals that are in solution in the water. In this reaction (shown in Eq 3-1), called hydrolysis, chlorine combines with water to form two compounds—hypochlorous acid (HOCl) and hydrochloric acid (HCl).

$$Cl_2 + H_2O \leftrightarrow HOCl + HCl \qquad (3\text{-}1)$$

Of the two compounds formed, the more important in the water treatment process is hypochlorous acid. This compound contains the active form of chlorine used in the oxidation, substitution, and disinfection reactions. This active form of the chlorine atom has a valence, or electrical charge, of +1. Chlorine can have several valences, as illustrated by Table 3-1.

Chemically, the more positive the valence, the stronger the oxidizing agent. This oxidizing strength, however, applies to its activity in chemical reactions and may not refer to its efficacy in water treatment, in which disinfection is the desired reaction. A higher valence for an element does not always mean that it will react with more chemicals in the water or at an increased rate. For example, although chlorine dioxide (CLO_2) has a higher valence (+4) and is a stronger oxidizing agent than hypochlorite (OCl^-), it does not react with ammonia compounds as the hypochlorite (or hypochlorous acid) form of chlorine does. The activity of the two forms of chlorine used in water treatment (hypochlorous acid/hypochlorite ion and chlorine dioxide) is such that their residuals are maintained throughout the water distribution system. Their presence provides an indication of water quality.

Table 3-1 identifies some of the more common compounds of chlorine and the chlorine valence in each of those compounds. It is characteristic of an oxidizing agent that it continues to react until no further oxidation is possible and the agent has been completely reduced or reached a state of stability.

Table 3-1

Valence states of the chlorine atom

Valence	Typical Chemical
− 1	Sodium chloride (table salt) (Cl^-)
+ 1	Sodium hypochlorite (liquid bleach) (OCl^-)
+ 2	Chlorine monoxide (ClO)
+ 3	Sodium chlorite (ClO_2^-)
+ 4	Chlorine dioxide (ClO_2)
+ 5	Sodium chlorate (ClO_3^-)
+ 7	Sodium perchlorate (ClO_4^-)

The chlorine that appears most frequently in nature is sodium chloride, which we know as ordinary table salt. The chlorine atom in salt is completely reduced and has a valence of –1. Thus, chlorine must acquire two electrons to change from the +1 valence of hypochlorous acid to the –1 valence of chloride.

The hydrolysis products of Eq 3-1 create an acid solution. The amount of pH depression in the solution depends on the concentration of chlorine in the water and the alkalinity of the water or the water's ability to buffer the acidity. Buffering is the action of salts dissolved in the water that allows water to resist changes in pH that might be caused by the addition of an acid or a base. Water often contains calcium and magnesium carbonates that react with acid to form a chemical salt that is in equilibrium with the acid and, as a result, maintains a constant pH. Hard waters with carbonate salts contribute to the water's ability to hold pH constant by buffering. Once the effect of the dissolved salts are neutralized by the addition of an acid, the pH will change rapidly with the subsequent addition of a chlorine water solution.

Hypochlorous acid dissociates or separates into two components, hydrogen ion (H^+) and hypochlorite ion (OCl^-) (shown in Eq 3-2).

$$HOCl \leftrightarrow H^+ + OCl^- \hspace{2cm} (3\text{-}2)$$

Like the hydrolysis reaction (Eq 3-1), the degree of dissociation of HOCl is primarily dependent on temperature and pH. An equilibrium reaction is pH dependent. This means that the equilibrium will shift to the right if the pH is raised and to the left if the pH is lowered.

A shift to the right increases the hypochlorite concentration. A shift to the left increases the hypochlorous concentration. The degree of dissociation and the equilibrium is further illustrated in Figure 3-1, which plots pH versus the reaction products. This graph plots pH on the horizontal axis, percent hypochlorous acid on the left vertical axis, and percent hypochlorite ion on the right vertical axis. The curve is a pictorial representation of Eq 3-2. A horizontal line drawn through the curve at any point will, at the points of intersection with each vertical axis, indicate the percent of the respective components in the solution. For example, at a pH of 7.2 and a temperature of 68°F (20°C), the solution would contain about 70 percent hypochlorous acid and 30 percent hypochlorite ion. At a higher pH of 8, the equilibrium shifts to about 21 percent hypochlorous acid and 79 percent hypochlorite ion. Notice that the sum of the two components will always total 100 percent.

Hypochlorous acid is defined as a weak acid. A weak acid is one with incomplete dissociation in comparison with strong acids, such as sulfuric (H_2SO_4) and nitric (HNO_3) acids, which are almost completely dissociated.

Figure 3-1

Hypochlorous
acid/
hypochlorite
distribution
versus pH

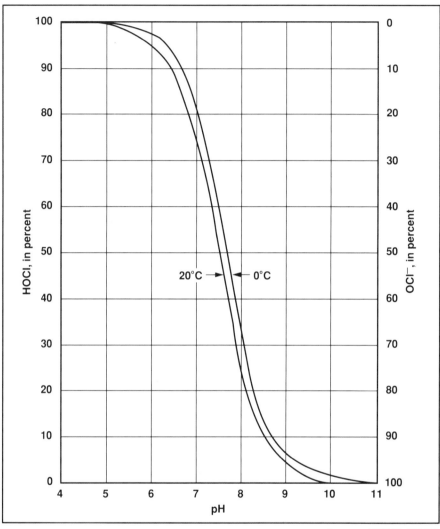

Source: AWWA (1973).

In the dissolution of chlorine in water, the pH of the resultant solution is affected by the amount of chlorine added and the alkalinity of the water. Generally, water treatment plant dosages of several mg/L do not change the pH of the water because most natural water systems are slightly alkaline and will buffer the chlorine addition. The exception is a water with low initial alkalinity. In general, it is recommended that the water alkalinity be determined so that the acidity caused by the addition of chlorine can be anticipated. Systems not experiencing any acid effect are "once-through systems" since the water to be treated has chlorine added only at one or two points during the treatment process. Chlorine dosages rarely exceed 5 or 10 mg/L. Each one part of chlorine added will neutralize the water's alkalinity. Typical reductions in alkalinity vary from 0.7 to 1.4 parts.

Chlorine used in recirculated systems can have a measurable effect on the pH of the treated water. The acids formed (HOCl and HCl) consume the alkalinity in the water. Chlorine is added to the same water repeatedly in a recirculated system. Only fresh makeup water is added to maintain the system at full levels. The subsequent reduction of the water's alkalinity and buffering ability allows pH depression to proceed. Examples of typical recirculated systems that experience this condition are swimming pools and industrial or commercial cooling water systems. These systems can accommodate for the lack of buffering ability developed in the water by using alkalies, such as caustic (NaOH) or soda ash (Na_2CO_3), to correct for the pH depression by chlorine gas. Such pH depression is not experienced in once-through water plants, wells, and similar facilities.

There are other chemicals that are sources of hypochlorous acid or hypochlorite ion used in water treatment. These include both sodium hypochlorite (NaOCl) (liquid bleach and javelle water) and calcium hypochlorite (Ca [OCl]$_2$) (powder bleach and chloride of lime). The former is widely used while the latter is used only occasionally. The reactions of these two compounds are shown in Eq 3-3 and 3-4.

$$NaOCl + H_2O \rightarrow HOCl + NaOH \qquad (3\text{-}3)$$

$$Ca(OCl)_2 + 2H_2O \rightarrow 2HOCl + Ca(OH)_2 \qquad (3\text{-}4)$$

The reaction products of both chemicals produce hypochlorous acid (as does chlorine gas), but the by-products of the two reactions are alkalies or bases and raise the solution pH, while chlorine gas lowers the pH. Solutions of both sodium hypochlorite and calcium hypochlorite have high pH values. Since the dosages of these solutions are rarely above 5 to 10 mg/L, the solution pH has little or no impact on the pH of the water, just as the acid pH of the chlorine gas solutions does not alter the pH of the treatment plant water. The exception, again, would be low-alkalinity waters. The impact that any solution, acid or base, would have is infinitesimal because the volume of the water is so large compared to the volume of the solution added. As mentioned, the only difficulty appears with recirculated water systems. In the case of either sodium or calcium hypochlorite addition, the pH will increase due to the increased alkalinity caused by the calcium and magnesium carbonates in the added hypochlorites. This is exactly the opposite of gas chlorination, which produces the acidic condition in recirculated waters.

A comparison of the oxidizing capability of the three forms of chlorine is illustrated in Table 3-2. Using chlorine gas as the benchmark, about 1.0 gal (3.785 L) of 12.5 percent sodium hypochlorite solution or 1.54 lb (699 g) of 65 percent calcium hypochlorite will give the same oxidizing capability as 1 lb (453 g) of chlorine gas.

Substitution Reaction

The reactions between chlorine and ammonia are called substitution reactions because the chlorine atom substitutes for the hydrogen atom in the ammonia molecule. The chemical equations for these reactions are complex. However, a simplified set of equations is most often used, as the complexities of the reactions have a negligible effect in water treatment. The equations defining the reactions are

$$HOCl + NH_3 \rightarrow NH_2Cl + H_2O \tag{3-5}$$

$$HOCl + NH_2Cl \rightarrow NHCl_2 + H_2O \tag{3-6}$$

$$HOCl + NHCl_2 \rightarrow NCl_3 + H_2O \tag{3-7}$$

The reaction products are classified as chloramines. The product in Eq 3-5 is called monochloramine, in Eq 3-6 dichloramine, and in Eq 3-7 trichloramine or nitrogen trichloride. In each of the compounds formed in the three equations, chlorine retains the +1 valence of the hypo-chlorous acid after substituting for or replacing the hydrogen atom. Thus, chlorine is still available as an oxidizing and disinfection agent. However, since chlorine has been combined with nitrogen and hydrogen in the ammonia molecule, it is not as readily available for oxidation and disinfection reactions. The chemical bonds holding chlorine in the molecule alter the availability, speed, and type of chemical reaction in which chlorine can be involved (University of California 1976).

Table 3-2

A comparison of oxidizing capabilities of three forms of chlorine

Material	Equivalent Oxidizing Capability
Chlorine gas	1 lb (454 g)
Sodium hypochlorite (12.5% by weight)	1.0 gal (3.785 L)
Calcium hypochlorite (65% strength)	1.54 lb (699 g)

Source: The Chlorine Institute (1995).

The degree of substitution or the number of chlorine atoms replacing hydrogen atoms in the chlorammoniation reaction is a function of pH, the relative amounts of ammonia nitrogen and chlorine present, and contact time. The initial substitution reaction to form monochloramine is rapid and depends primarily on adequate mixing to bring the reactants together. The pH, temperature, and chlorine–ammonia ratio govern whether and how many mono-, di-, and trichloramines are formed. The chloramine compounds are also called *combined chlorines*, as the chlorine that is combined with ammonia are not as available for disinfection. Hypochlorous acid and hypochlorite ion are called *free chlorine* because these compounds are free to disinfect. Together, the two groups are referred to as *total chlorine*.

Characteristics of the Forms of Chlorine

Free chlorine (hypochlorous acid and hypochlorite ion) and combined chlorine (mono-, di-, and trichloramine) have different chemical and physical characteristics. While free chlorine is far less volatile than combined chlorines, it is not easily removed by aeration. However, combined chlorines are easily removed by aeration, and the volatility or vapor pressure increases with increases in chlorine atoms in the combined chlorine compounds. Monochloramine is least volatile and has the lowest vapor pressure. Dichloramine is more volatile and has a higher vapor pressure than monochloramine. Trichloramine has the highest vapor pressure and is the most volatile. Tables 3-3 and 3-4 provide relative taste-and-odor values of the chloramine family of compounds.

With the increasing volatility there is a concurrent increase in odor characterized as chlorinous in nature. The odor level increases with the increase of chlorine atoms in the chloramine compounds.

The taste and odor of finished, treated water can also be drastically affected by the presence of chloramines. Di- and trichloramines contribute more significantly to tastes and odors than free chlorine and monochloramine. As in odors, tastes are more drastically affected with the increase in chlorine atoms substituted. Free chlorine and monochloramine in treated water contribute minimally to the taste of water.

Compound	Level (mg/L)
NH_2Cl	0.48
$NHCl_2$	0.13
NCl_3	0.02

Table 3-3

Odor threshold of chloramines

Source: Krasner and Barrett (1984).

Table 3-4

Taste threshold
of chloramines

Compound	Level (mg/L)
NH_2Cl	0.65
$NHCl_2$	0.15
NCl_3	0.02

Source: Krasner and Barrett (1984).

Tastes appearing after the water has left the treatment plant can usually be attributed to reactions of free chlorine with growths in the distribution system, which produce chloramines or other taste-and-odor formers.

Combined chorines are less aggressive, more persistent, and react more slowly with oxidizable materials and bacteria in water than free chlorine. This characteristic of combined chorines contributes to its use in water treatment as a means of trihalomethane (THM) inhibition.

The distribution of mono- and dichloramines as a function of pH is illustrated in Figure 3-2 in a curve that has a remarkably similar shape to the free chlorine curve of Figure 3-1. Knowledge of this curve and control of the chlorine-to-ammonia ratio enable the operator to control the reaction to limit the formation of certain forms of chloramines (see breakpoint reaction on page 33).

Inorganic Oxidation Reactions

Oxidation of soluble iron, manganese, and sulfides in water is a common use of chlorine in water treatment. The oxidized iron, manganese, and sulfides form insoluble compounds. Filters are used to remove these oxidation products. In some cases, the oxidation product may be a soluble innocuous ion, such as sulfate (produced by oxidation of the sulfide ion). The oxidation of these and other inorganic chemicals (e.g., nitrite) are shown in Eq 3-8 through 3-12.

$$HOCl + Fe^{+2} \rightarrow Fe^{+3} \qquad (3\text{-}8)$$

$$HOCl + Mn^{+2} \rightarrow Mn^{+4} \qquad (3\text{-}9)$$

$$HOCl + S^{-2} \rightarrow S \qquad (3\text{-}10)$$

$$HOCl + S \rightarrow SO_4^{-2} \qquad (3\text{-}11)$$

$$HOCl + NO_2^- \rightarrow NO_3^- \qquad (3\text{-}12)$$

The amount of chlorine required in these inorganic reactions and the effect on water alkalinity are summarized in Table 3-5.

Iron and manganese will appear at points of use in the system and leave undesirable stains if not oxidized and filtered. Sulfides, if not oxidized and removed, will produce an undesirable rotten-egg odor that

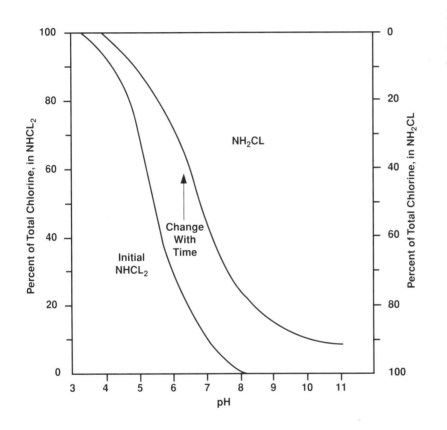

Figure 3-2

Distribution of mono- and dichloramine

Inorganic Reactant	Part of Chlorine Required per Part of Inorganic	Alkalinity Consumed per Part of Chlorine Added
Iron	0.6	0.9
Manganese	1.3	1.5
Nitrite	1.5	1.8
Sulfide to sulfur	2.1	2.6
Sulfur to sulfate	6.2	7.4

Table 3-5

Stoichiometric chlorine requirements and alkalinity consumption with inorganics

Source: AWWA (1973).

31

makes the water unpalatable. Chlorine, when oxidizing iron and manganese, produces insoluble ferric and manganic compounds that must be removed by coagulation and filtration. With sulfides, there are two reaction steps possible. The first reaction, shown in Eq 3-10, · produces colloidal sulfur that must be removed by filtration. The second reaction, shown in Eq 3-11, produces a soluble sulfate ion that has no taste or odor, is not considered objectionable, is soluble, and is not removed from the finished water.

Nitrites are soluble compounds that are oxidized to nitrates, also soluble materials. Nitrites are compounds that can produce adverse responses in the human system, such as affecting the hemoglobin in blood and the ability to transfer oxygen (Snoeyink and Jenkins 1980). Nitrates affect the human body in a similar fashion (Kleinjans 1993).

Organic Oxidation Reactions

Reactions of chlorine with some organics produce disinfection by-products (DBPs). The most common are THMs. THMs include such compounds as chloroform, bromodichloromethane, dibromochloromethane, and bromoform. These compounds are classified as probable human carcinogens.

The current maximum permissible THM level in drinking water is 0.1 mg/L (100 μg/L) or 100 ppb. At the time of this writing, the US Environmental Protection Agency (USEPA) was reviewing the toxicity of this class of compounds to establish maximum allowable levels for drinking water. The level that appears likely to be established for THMs is 0.08 mg/L (80 μg/L), or 80 ppb. It's important to be aware of the current permissible levels, since these levels are continuously being reevaluated in light of new health information.

There are other products of chlorine–organic reactions that have been recently determined to be probable cancer-causing agents, but not of the THM group. Table 3-6 is a list of some of the generally recognized DBPs, THMs, and other compounds commonly referred to as DBPs.

The organics associated with THM production are those from the humic and fulvic acids. There are many factors influencing the formation of DBPs. Among them are contact time (CT), temperature, pH, precursor type and concentration, disinfectant type and concentration, ratio of oxidant to precursor, and concentrations of bromide and nitrogen.

Chlorine has been used to reduce taste, color, and odor in water systems. Current water treatment regulations limit the use of chlorine for this purpose, however, because of the possible production of undesirable DBPs. Generally, the use of a pretreatment method that will remove DBP precursors or inhibit the formation of DBPs is the water treatment method of choice. These methods may include one or more of the following: membrane filtration, pretreatment or preoxidation with ozone

Table 3-6
Chlorination
disinfection
by-products

Trihalomethanes	Haloacetic Acids
chloroform	monochloroacetic acid
bromodichloromethane	dichloroacetic acid
dibromochloromethane	trichloroacetic acid
bromoform	monobromoacetic acid
	dibromoacetic acid
	tribromoacetic acid
	bromochloroacetic acid
	bromodichloroacetic acid
	dibromochloroacetic acid
Haloacetonitriles	**Others**
dichloroacetonitrile	chloral hydrate
trichloroacetonitrile	haloketones
bromochloroacetonitrile	
dibromoacetonitrile	
tribromoacetonitrile	
Cyanogen Halides	
Halopicrins	

Source: Fielding et al. (1993).

or chlorine dioxide, chloramination, reduction in chlorine dosage, or
relocation of the points of chlorination.

The reaction rates with organic compounds are slower than those
with inorganic compounds. Sometimes they can be difficult to quantify
since the rate, demand, and end products may vary considerably. In
addition, the nature and complexity of the organic molecules make it
difficult to quantitatively predict the reaction products. The determination of chlorine demand is usually established by laboratory testing
or jar testing to determine the desired treatment level. These
requirements typically change seasonally in surface water treatment
plants. The complexity of new regulations may require pilot plant
evaluation to effectively determine the best overall treatment and the
most effective role for chlorine. When chlorine is used in the presence of
organics, the reaction products are of concern and the results should be
reviewed thoroughly to determine the end products.

The Breakpoint Reaction

A phenomenon occurs with chlorine reactions in water containing
ammonia that deserves special attention. The phenomenon is called the
breakpoint reaction. The reactions of chlorine with ammonia are cited
earlier in this chapter. The chloramines formed are weak disinfectants

and weak oxidants compared to free chlorine. When virus inactivation and manganese oxidation are required, chloramines cannot be used effectively. The breakpoint phenomenon and the use of the breakpoint process results in the oxidation of ammonia and the elimination of chloramines, a significant contributor to the chlorinous taste-and-odor problem.

Reactions between chlorine and ammonia can take many paths but depend primarily on the chlorine–ammonia ratio and pH. The breakpoint reaction rate is affected by initial concentrations, pH, temperature, and other factors. The continued addition of chlorine to water containing ammonia first forms monochloramine, then converts to dichloramine, and finally to trichloramine. At some point, the addition of more chlorine will start to break down the chloramines. At even higher concentrations, decomposition to nitrogen gas and nitrate is probable.

The exact mechanism of chloramine oxidation has not been established with unanimous agreement. Some of the more common equations that can represent this oxidation are shown in Eq 3-13, 3-14, and 3-15.

$$NHCl_2 \;\rightarrow\; NOH \;\rightarrow\; H_2N_2O_2 \;\rightarrow\; N_2O \qquad (3\text{-}13)$$

$$NHCl_2 + NOH \;\rightarrow\; H_2 \qquad (3\text{-}14)$$

$$H_2N_2O_2 + HOCl \;\rightarrow\; NO \qquad (3\text{-}15)$$

The curve that illustrates the reaction is called the *breakpoint curve* (Figure 3-3). This curve plots the chlorine dosage in mg/L along the horizontal axis and the total (free and combined) chlorine residual in mg/L along the vertical axis. The line plotted at a 45° angle is called the *zero demand line* and represents the ideal situation in which all chlorine added to the water is measured as a free chlorine residual in the water. In zero demand water there is no loss of chlorine addition due to chlorine demand (University of California 1976).

The presence of reducing agents and other inorganic chlorine-demanding compounds will consume any initial chlorine dosage, which results in zero chlorine residual until this demand is satisfied. This section is identified as zone 1 in Figure 3-3. Once this initial demand is met in the presence of ammonia, any additional dosage will be measured as a combined chlorine residual. This residual will increase in proportion to the increase in dosage and is identified as zone 2.

At some point, the dosage/residual relationship will cease to be linear and the amount of combined and total residual increase will flatten out as shown in zone 3. Subsequent increases in chlorine dosage will cause a decrease in the chlorine residual as seen in zone 4. This continuing decrease of total residual will taper off and the chlorine

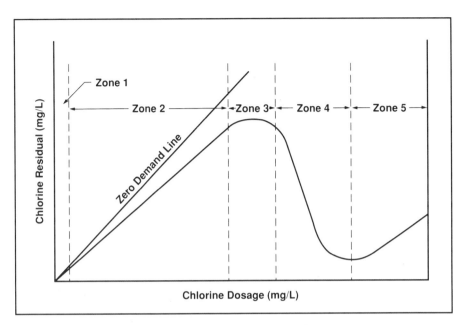

Figure 3-3
Breakpoint
curve

residual will reach a minimum point after which it will once again increase with continued chlorine addition. This minimum point is called the breakpoint and represents that point in the treatment process where all ammonia compounds have been consumed. Any further chlorine addition will increase the residual chlorine as free chlorine identified in zone 5. Combined chlorines predominate in zones 2, 3, and 4, while free chlorine predominates in zone 5.

Some waters treated by the breakpoint process may never reach a zero residual since free chlorine may have formed in zone 4 and, depending on the water, organic nitrogen compounds present resist the oxidation by chlorine. This organic nitrogen is sometimes referred to as an irreducible minimum.

Stoichiometrically, the chlorine/ammonia reaction has been established at a minimum of 7.6 parts of chlorine per part of ammonia nitrogen. In practice, the breakpoint reaction is considered complete when the chlorine dosage has reached 10 parts of chlorine to 1 part of ammonia nitrogen with the pH maintained in the 7 to 8 range. The free residual at the time breakpoint is reached will be about 80 percent of the total residual. The free or combined chlorine percentage of the total chlorine will vary depending on the water supply, the season, water temperature, and pH. Each water plant must determine its own breakpoint curve. Analysis of the water for the various forms of chlorine at different chlorine dosages allows for determination of the particular breakpoint curve for the water. Procedures from *Standard Methods for the*

Examination of Water and Wastewater (AWWA 1995) should be followed for these determinations.

The Swimming Pool

An interesting study in chlorine–water chemistry is the situation that occurs in the treatment of swimming pool water. In a survey conducted by The Chlorine Institute to determine the public's perception and knowledge of chlorine and its uses, it was found that people only recognized chlorine as a useful chemical for disinfection of swimming pool water and drinking water (The Chlorine Institute 1991). Significantly, chlorine's use as a swimming pool treatment was positioned equal to or greater than the use of chlorine in water treatment. Because of the unique conditions that occur in a swimming pool, e.g., recirculated system, high ammonia content, and attention to pH, the chemistry associated with pool water treatment is worth reviewing. Much of what occurs in pool water treatment with chlorine is relevant to drinking water treatment. Moreover, the swimming pool represents a unique opportunity for further comment on the breakpoint process.

Pool water receives a great many nitrogen compounds in the form of perspiration and urine. From these materials, urea is hydrolyzed to form ammonia compounds. The presence of ammonia in the water provides a clear path for chloramine formation and the development of chlorinous odors. Water in an improperly understood and poorly treated pool can lead to chlorine odors and stinging of the eyes. This condition is more easily observed at indoor pools and at the water surface in outdoor pools. The chlorine odor and eye stinging are often attributed to overchlorination.

In actuality, chlorine odor in pools is a symptom of inadequate chlorine addition and/or pH control. Effective disinfection can only be achieved with free chlorine. Unless the pH is above 6.5, breakpoint will not occur and nitrogen chloride will be formed. With high ammonia concentrations, high chloramine concentrations can be formed and strong odors will result. The chlorine dosage may indirectly be reduced. This generally does not alleviate the odor condition and may affect disinfection.

The proper course of action is to increase the chlorine feed rate and chlorine dosage to reduce the combined chlorines, and to operate the pool in the free chlorine residual range. Pool odor is a classic example of improper treatment of water with chlorine and demonstrates a misunderstanding of the reactions of chlorine with ammonia compounds. Similarly, in water distribution systems, isolated or distant service connections may well experience similar chlorinous tastes and odors that could be mistakenly attributed to excess chlorine. Analysis of the water to determine the free and combined residual level is important. If the analysis shows that combined chlorines predominate, then increased chlorine dosage is the solution to ensure the presence of free chlorine, just as in the swimming pool example.

Disinfection

Disinfection of water refers to the removal or inactivation of pathogenic organisms. Disinfection does not mean sterilization. The result of effective disinfection is the production of potable, or drinkable, water. Chlorine has long been recognized as a very effective and efficient disinfection agent.

It is important to discuss the process of disinfection to provide some understanding of the steps in the treatment process. All bacteria have the same essential composition. They contain some or all of the following components: a means of food ingestion, food absorption, waste discharge, mobility, and ability to replicate. All bacteria are single-celled and have a slime coating on the exterior of the cell wall. A typical organism is illustrated in Figure 3-4. The exterior coating of these organisms has a negative charge.

In the process of disinfection, the disinfectant attempts to disrupt the normal life processes of the organism. This is done by penetrating the cell wall of the organism and upsetting the natural life cycle processes or altering the enzymes. With the cycle so disrupted, either the organism dies or the species cannot reproduce and the water is made bacteriologically safe.

The electrical charge on the surface of the organism and the corresponding charge of the disinfectant is important in the process. Various disinfectant forms have electrical charges that affect their ability to disinfect. The hypochlorous acid produced on addition of chlorine to water has no charge. The dissociation product, hypochlorite ion, has a negative charge. The chloramines all have a neutral charge. The ability of a disinfectant to function effectively is influenced by these charges. Since the outer coating of a cell has a negative charge, any disinfectant with a neutral or positive charge will penetrate the slime coating more rapidly because the resistance is less. The hypochlorite ion, negatively charged, meets more resistance than the hypochlorous acid, neutrally charged, and takes longer to disinfect. In addition, the size of the molecule will affect its ability to penetrate the slime coating. Hypochlorous acid is a less complex molecule and, as such, penetrates the cell much more rapidly than the chloramines, more complex molecules, which also have a neutral charge.

The speed of the disinfection reaction is a concern in water treatment. There are some processes that affect the speed of this reaction. A good example is disinfection at water treatment plants using the lime-softening process. Lime softening increases the pH of the water to

Figure 3-4

Typical organism

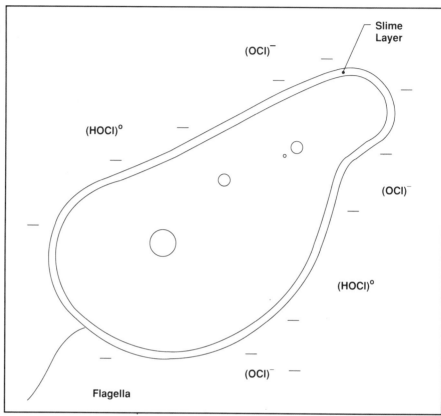

aid in the removal of calcium and magnesium salts. In the process, the pH of the water will be 10 or more. Any attempt to disinfect water at a pH of 10 or more will not be very effective. Any chlorine added to the water at that pH will be present in the hypochlorite form. To be effective, disinfection with chlorine is best conducted at pH levels below 7.5. At this pH, hypochlorous acid predominates and cell penetration is more rapid. Although such other factors as temperature, time, and the type of organism may affect the efficacy and speed of the disinfection process, none is more important than pH, particularly when chlorine is the disinfectant. Most state regulations require a minimum contact time (CT) of 30 min with the disinfectant before the first service connection may be made or water pumped into the distribution system. Since the life cycle of most organisms is less than 30 min, effective kill with chlorine can be expected as long as the pH is less than 8.

$C \times T$ **Disinfection Concept**

The water industry has for some time recognized the need for adequate exposure to the disinfectant and sufficient disinfectant dosage. In the 1980s the combination of the two was brought into focus with the development of the $C \times T$ values for the various disinfectants.

$C \times T$ represents the combination of dosage of the disinfectant (C) and the length of time (T) water has been exposed to a minimum amount of the disinfectant. Mathematically it is represented in Eq 3-16.

$$C \times T = \text{concentration} \times \text{time} \qquad (3\text{-}16)$$

Where:

concentration = final disinfectant concentration, in mg/L
time = minimum exposure time, in minutes

Two particular types of organisms have been chosen as disinfection surrogates to aid in assessing disinfection effectiveness. They are *Giardia* and viruses. $C \times T$ values for the disinfection of surface waters have been set at a three log (99.9 percent) reduction in *Giardia* and a four log (99.99 percent) reduction in viruses. Temperature and pH have an impact on the $C \times T$ value, as shown in Tables 4-1 and 4-2.

It is important to recognize that the use of chlorine as the disinfectant is only one part of the treatment process. In fact, the USEPA acknowledges this in the drinking water regulations evolving under the Safe Drinking Water Act. Of equal importance is the need for improved filtration or removal of organisms as well as inactivation with chlorine or other disinfectants. A combination of proper disinfection and improved removal by filtration is often most effective in providing a disinfected water. Recent outbreaks of cryptosporidiosis may result in revised goals, primarily in the area of filtration. Chlorination is not recognized as an effective means for *Cryptosporidium* removal, although some research has indicated a possible inactivation using a chlorine/chloramine treatment (Finch 1996). Currently filtration is the most appropriate removal technique.

Reduction Reactions

Where an oxidizing agent gains electrons in a reaction, there must be a reactant that loses electrons. Chemicals that lose or provide the electrons are called *reducing agents*. Reducing agents seek to provide electrons.

Therefore, in Eq 3-8 through 3-12, iron, manganese, and sulfide (the reactants with chlorine) are classified as reducing agents because the reaction products have all donated electrons and their valences, or charges, have increased in positive value. Iron changed from +2 to +3, manganese from +2 to +4, and sulfide from –2 to 0 and then to +6. In these equations, the reactions were classified as oxidation, although the oxidized participants in the reaction were reduced.

There are instances when the reverse is desired, that is, the elimination of an oxidant. Reducing agents most frequently used are those involved with sulfur, predominantly sulfur dioxide (SO_2), and the sulfur salt solutions, including sodium bisulfite ($NaHSO_3$) and sodium sulfite (Na_2SO_3). Though not used frequently today, sodium thiosulfate ($Na_2S_2O_3$) has been used as a reducing agent in some water treatment plants. Equations 3-17 through 3-20 define the use of these reducing agents with chlorine.

$$SO_2 + Cl_2 + 2H_2O \rightarrow 2HCl + H_2SO_4 \tag{3-17}$$

$$NaHSO_3 + Cl_2 + H_2O \rightarrow NaHSO_4 + 2HCl \tag{3-18}$$

$$Na_2SO_3 + Cl_2 + H_2O \rightarrow Na_2SO_4 + 2HCl \tag{3-19}$$

$$2Na_2S_2O_3 + Cl_2 \rightarrow Na_2S_4O_6 + 2NaCl \tag{3-20}$$

One might ask, "Since chlorine appears in the equations, why not call these reactions oxidations?" The reason is that the desired result was to eliminate the chorine; thus, the reaction is defined as one of reduction. The most common application of the reduction reaction in water treatment is in dechlorination or removal of excess chlorine.

References

American Public Health Association, American Water Works Association, and Water Environment Federation. 1995. *Standard Methods for the Examination of Water and Wastewater.* 19th ed. Washington, D.C.: APHA.

American Water Works Association. 1973. *AWWA Manual M20, Water Chlorination Principles and Practices.* Denver, Colo.: AWWA.

Fielding, M., N. Mole, P. Jackson, and H. Harth. 1993. *Review of Chlorination and Other Disinfection By-products and Implications for the U.K. Water Industry.* Report UC 1886. Water Research Centre, U.K.

Finch, G.R., L.L. Guyrek, and M. Belosovic. Unpublished. The Effect of Chlorine on Waterborne *Cryptosporidium parvum.* University of Alberta, Edmonton, Alta.

Kleinjans, J.C.S., H.J. Albering, A. Mark, J.M.S. van Maanen, and B. van Angen. 1993. *Environmental Health Perspectives,* Vol. 94, pp 189–193. Maastricht, Netherlands: Limburg Univ.

Krazner, S.W., and S.E. Barrett. 1984. *Aroma and Flavor Characteristics of Free Chlorine and Chloramines.* In Proceedings of 12th Annual AWWA WQTC. Denver, Colo.: AWWA.

Snoeyink, V.L, and D. Jenkins. 1990. *Water Quality and Treatment*. 4th ed. Denver, Colo.: AWWA.

The Chlorine Institute. 1991. *Study of the Status of Chlorine in the Mind of the Average Citizen*. Washington, D.C.: The Chlorine Institute.

———. 1995. *Pamphlet 65, Personal Protective Equipment for Chlorine and Sodium Hydroxide*. Washington, D.C.: The Chlorine Institute.

University of California. 1976. *Kinetics of Breakpoint Chlorination and Disinfection*. Report 76-2, Berkeley, Calif.: Sanitary Engineering Research Laboratory, University of California.

Process Applications

This chapter examines various processes and procedures involving chlorine and ammonia in the treatment plant and distribution system. First, a few definitions are established and the objectives of each step in the treatment process identified.

Types of Chlorine Residual

Chlorine exists in several forms in water. When dissolved in water, chlorine reacts to form hypochlorous acid. Depending on the pH, the hypochlorous acid dissociates to form hypochlorite ion. Both hypochlorous acid and hypochlorite ion are called *free chlorine* or *free available chlorine* (FAC). The sum of their values is equal to the free chlorine content of the water.

The products of the reaction between chlorine and ammonia are the chloramines. All three of these forms are called *combined chlorine*. The sum of their values is equal to the combined chlorine content. Free plus combined chlorine is referred to as *total chlorine*.

$$\text{Free Chlorine } + \text{ Combined Chlorine } = \text{ Total Chlorine} \qquad (4\text{-}1)$$

Where:

Free chlorine $= \text{HOCl or OCl}^-$
Combined chlorine $= \text{NH}_2\text{Cl}, \text{NHCl}_2, \text{or NCl}_3$

43

In either of these forms, chlorine is available as an oxidant or disinfectant in the water treatment processes. Free chlorine is far more aggressive in water treatment than combined chlorine. Free chlorine reacts more rapidly with organics and inorganics and, therefore, does not persist as long as combined chlorines. On the other hand, combined chlorine is a weaker oxidizing agent. It reacts more slowly than free chlorine and persists for a longer time. As a result, free chlorine is often used as the primary disinfectant and combined chlorine as a secondary disinfectant. Combined chlorine is also used to provide a residual in long distribution systems. A combined residual at the plant clearwell will have a far better chance of being present at the end of the distribution system than the free residual. The use of combined chlorine in the distribution system also stops the formation of trihalomethanes (THMs).

Chloramination

In the chloramination process as it is applied to the treatment of water, monochloramines (containing one chlorine atom) are preferred. The formation of the di- and trichloramine species is minimized by controlling the chlorine and ammonia ratios (3 or 4:1) and the pH (7 to 8). The di- and trichloramines will contribute to the taste and odor of the treated water if they are allowed to develop. The use of monochloramine is acceptable because it contributes least to drinking water taste-and-odor problems. Chlorine is usually added after the addition of ammonia, although the reverse is also practiced.

A note of caution is in order for the water treatment plant that uses or is about to use chloramines. Chloramines present in the water at the consumer's point of use will damage kidney machines and animal species in aquaria. Utilities using chloramines must notify the consumers at risk so that corrective action may be taken (e.g., the use of filters, aeration, and so forth) (Kirmeyer et al. 1993).

Surface Water Treatment

The Safe Drinking Water Act has had a profound effect on the treatment of water, specifically the treatment of surface water. Forever altered are the treatment practices, techniques, and principles that have evolved over years of use. By imposing requirements on the quality of drinking water, the law has changed these treatment practices and techniques and forced the evaluation of new treatment methods to achieve better results and meet more stringent regulations. Discussions are ongoing in the water industry as to improved methods and practices in this regard.

Historically, the surface water plant would provide chlorination at the raw water intake or flash mixer and in the clearwell prior to the water's entry into the distribution system. In addition, there may have been other intermediate chlorination points (e.g., filters) used from time to time. The development of the $C \times T$ concept, its adoption by the US Environmental Protection Agency (USEPA) and the recognition of the disinfection by-products (DBPs) from chlorine altered the free-wheeling chlorination approach forever.

The $C \times T$ Concept

In the 1980s the $C \times T$ concept was introduced based on the work of several research efforts. As mentioned in chapter 3, the $C \times T$ value is used to indicate the effectiveness of disinfection with chemical oxidants. The use of $C \times T$ is one more design tool available and helps establish more structured operating criteria for water treatment disinfection (USEPA 1989).

$C \times T$ is the product of the disinfectant residual and exposure time in the water to reach certain preestablished levels of disinfection. (For additional discussion, see chapter 3.) The water treatment plant operator can alter either the dosage or the treatment time to reach the prescribed $C \times T$ value. $C \times T$ values are tabulated for free chlorine and combined chlorine in Tables 4-1 and 4-2. Table 4-1 represents the $C \times T$ values for *Giardia lamblia* and Table 4-2 the values for viruses (USEPA 1989). For comparison with free chlorine and combined chlorine, both tables include $C \times T$ values for other disinfectants, such as ozone and chlorine dioxide, that are gaining wider acceptance in water treatment due to the THM concern. Ozone and chlorine dioxide are considered primary disinfectants. Preoxidation with ozone or chlorine dioxide are methods of minimizing THM formation.

The factors that have the greatest influence on the $C \times T$ value are temperature and, in the case of chlorine, pH. From Table 4-1 for *Giardia lamblia*, it can be seen that pH has little influence on the action of chlorine dioxide, ozone, and combined chlorine, while free chlorine is decidedly influenced. This is to be expected, since the higher pH shifts the dissociation equilibrium toward hypochlorite ion and the disinfection effectiveness is reduced. Temperature will affect all four options. The higher the water temperature, the lower the $C \times T$ value required. From Table 4-2 for viruses, it can be seen that both pH and temperature have a clearly defined influence with all four forms of disinfection. In general, it can be said that the warmer the water and the lower the pH, the more effective the disinfection by free chlorine. The $C \times T$ values for chloramines are about 20 times larger than free chlorine. For this reason, chloramines require a far greater dosage and/or considerably longer treatment time than free chlorine to reach equivalent disinfection levels.

In the determination of $C \times T$ values, the contact time used in calculating $C \times T$ is considered to be the detention time that is equaled or

Table 4-1

C × T values for 99.9 percent reduction of *Giardia lamblia*

Disinfectant	pH	Temperature, °F (°C)					
		33.8 (1)	41 (5)	50 (10)	59 (15)	68 (20)	77 (25)
Free chlorine	6	165	116	87	58	44	29
	7	236	165	124	83	62	41
	8	346	243	182	122	91	61
	9	500	353	265	177	132	88
Ozone	6–9	2.9	1.9	1.4	0.95	0.72	0.48
Chlorine dioxide	6–9	63	26	23	19	15	11
Chloramines	6–9	3,800	2,200	1,850	1,500	1,100	750

Source: USEPA (1989).

Table 4-2

C × T values for inactivation of virus, pH 6–9

Disinfectant	Log Inactivation	Temperature, °F (°C)					
		39.9 0.5	41 (5)	50 (10)	59 (15)	68 (20)	77 (25)
Free chlorine	2	6	4	3	2	·1	1
	3	9	6	4	3	2	1
	4	12	8	6	4	3	2
Ozone	2	0.9	0.6	0.5	0.3	0.25	0.15
	3	1.4	0.9	0.8	0.5	0.4	0.25
	4	1.8	1.2	1.0	0.6	0.5	0.3
Chlorine dioxide	2	8.4	5.6	4.2	2.8	2.1	—
	3	25.6	17.1	12.8	8.6	6.4	—
	4	50.1	33.4	25.1	16.7	12.5	—
Chloramines	2	1,243	857	643	428	321	214
	3	2,063	1,423	1,067	712	534	356
	4	2,883	1,988	1,491	994	746	497

Source: USEPA (1989).

exceeded by 90 percent of the fluid passing through the system. This time is referred to as the t_{10} time.

Tables 4-1 and 4-2 represent the governing criteria in the design and operation of the water treatment plant. $C \times T$ credit is usually obtained from the point of first addition to the point of lowest residual measurement in the treatment plant. Prefiltration credit may not be allowed in some states, and future $C \times T$ credit may be determined after coagulation and filtration have been achieved.

$C \times T$ is calculated using the lowest continuous residual and the exposure time for that residual. Judicious location of the point of addition can either reduce the dosage or take advantage of disinfectant residence in nontraditional points in the water treatment process to obtain $C \times T$ credit. Plant operating personnel should recognize the plant $C \times T$ values and operate to obtain optimum results. Methods to obtain optimum results may include a reduction in dosage or the addition of chlorine at different locations, both of which would help maintain the necessary $C \times T$ value. Discussions with the engineering consultants that designed the facility will provide the information necessary and the choices available. In some situations, storage is now considered as a treatment basis for disinfection.

Each treatment plant should have its set of $C \times T$ values that represent residence times at different flow rates, water temperatures, addition points, and dosages for both free and combined chlorine.

Concern over DBPs may cause plant operators to consider the formation of THMs as the prime concern in plant operation. Of greater importance is operating the water treatment plant to maintain the proper $C \times T$ value so that disinfection is attained. Variations in demand and plant water flow must always be considered when determining the desired chlorine or ammonia feed rate. The economy of the plant operation is clearly spelled out in the consumption of chemicals (i.e., chlorine and ammonia). There are clear advantages to using a continuous chlorine residual analyzer in conjunction with a water flow signal to both monitor and control the chlorine and chloramination process. The use of proper equipment can make just as much an impact as the concern for THMs and $C \times T$ values.

Points of Application

Groundwater

Chlorination application guidelines differ for groundwater sources and surface water sources. In general, groundwaters are of a good quality, exhibit little or no turbidity, and require only chlorination to meet drinking water standards. Chlorine dosages of 1 to 2 mg/L are usually sufficient to achieve disinfection and to provide the desired residual. There are exceptions, depending on the source of the water. Organic contaminants can cause taste, odor, or color problems that would require additional treatment (e.g., ozonation and filtration). These are found in wells that obtain water from aquifers located in prehistoric swamps or from shallow wells near swampy areas or rivers with high organic loads.

Surface Waters

In surface water treatment plants, common points of chlorine application include the raw water intake, flash mixer, filters, and clearwell. Other locations may be used from time to time depending on the plant process train. For example, in lime-softening plants, chlorine should not be added until the pH is returned to the 7 to 8 range. This is after calcium and magnesium salts have been removed in the clarifiers, settling basins, and recarbonation process. Treatment with chlorine after the lime addition (under high pH conditions) could result in insufficient kill, would call for higher doses of chlorine, and may not meet the required $C \times T$ value.

The location of the points of addition and the quantities of chlorine added must always be evaluated for THM formation potential. Whereas plants at one time were operated with little concern about the impact of the formation of these by-products, today's operating personnel must keep DBPs in mind and add only enough chlorine to achieve the result desired while minimizing the formation of undesirable by-products.

Prechlorination

Prechlorination is the addition of chlorine to the water prior to any of the other plant processes. Facilities may add ammonia before, after, or with chlorine addition to form chloramines. Chloramines reduce the DBP formation potential.

Prechlorination can take place in the flash mixer or at the raw water intake structure. The latter point of addition is being considered more and more often by utilities, particularly those in the eastern part of the country, due to the zebra mussel phenomenon that began in the late 1980s.

Zebra Mussels

Zebra mussels are indigenous to the fresh waters of Europe. It is believed that, sometime during the 1980s, ships arriving in the Great Lakes from eastern European ports pumped bilge or ballast water containing zebra mussel larvae and/or adult mussels from the waters of their European sources. From that modest beginning the animals rapidly adapted to the waters of the Great Lakes and spread into their tributaries. The mussels reached the Mississippi and Ohio rivers by attaching themselves to boats or by being carried by currents to the remote locations of these rivers. There have been many reported sightings in seemingly unrelated locations in the West, Southwest, East Coast, and Southeast. Figure 4-1 shows reported sighting locations as of 1994.

Movement to these unconnected waters has apparently been accomplished by waterfowl migration or attachment of the larvae form,

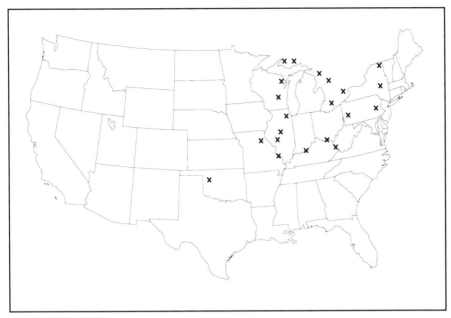

Figure 4-1

Locations of reported zebra mussel sightings

called veligers, to the many pleasure boats that are moved frequently from lake to lake and river to river. Some utilities are initiating quarantines on new pleasure craft arriving at their lakes or reservoirs. The boats must sit in dry-dock isolation for a period of time so that the animals will die off. Only after that minimum time will the boats be allowed into the water. If the owner removes his vessel from the quarantine, then the process will start over again.

Zebra mussels create problems for water treatment plant operations by attaching themselves to the intake structure, piping, bar screens, and other components of the plant water system. Effective treatment of zebra mussels is achieved by the addition of chemical oxidants to the intake structure. One of the most effective oxidants is chlorine. Usually this treatment is accomplished by shock treatment using intermittent dosages of several parts per million (ppm) of chlorine (Figure 4-2).

Shock treatment is the addition of higher than normal dosages of chlorine for shorter periods of time. This method results in lower total chlorine consumption and the ability to effect a more complete and rapid kill. Shock treatment also results in lower THM formation potential. Since zebra mussels spawn during the times of the year when the water is warmer, shock treatment during the winter is generally not practiced. Dosages of 2 to 5 mg/L of chlorine for short time periods are often effective against zebra mussels. Shock dosages of 10 mg/L have also been employed for periods of 30 min and have been found to be effective.

Shock treatment may be the most satisfactory technique to combat zebra mussels, but care must be taken that this addition does not cause the undesirable formation of DBPs. A fine line is drawn for the operator to maintain plant operations while watching the results in the finished water.

It may be advantageous or necessary to use other forms of treatment, such as ozone, chlorine dioxide, or potassium permanganate to kill zebra mussels. Although bromine compounds, such as bromine chloride and activated bromide salts, have been used in zebra mussel abatement, their use at water treatment intakes is not recommended. The presence of bromide enhances DBP formation when chlorine is added.

Each site has specific conditions that must be evaluated in regard to the zebra mussel problem. Operating personnel must analyze their local situations to determine the most practical and cost-effective means of resolution.

Intermediate Chlorination

Intermediate chlorination is the addition of chlorine at intermediate points throughout the plant or between various treatment processes. The addition of chlorine ahead of the filters is frequently practiced. Chlorine added here may enhance filter performance by minimizing the biological buildup on filter beds and lengthening the time between filter cleanings. Chlorine may be added on an intermittent, as needed, or continuous basis. Continuous addition can cause additional DBP formation because of the increased chlorine contact time (CT). Although ammonia can be added in conjunction with chlorine, the use of free chlorine is normally far more effective in maintaining the filter operation at an optimum level.

Figure 4-2
Shock treatment for zebra mussel eradication

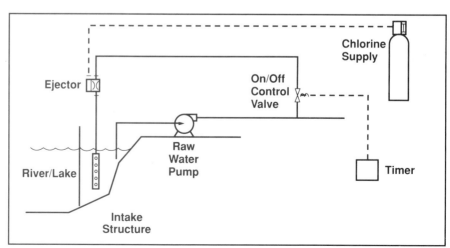

Postchlorination/Postammoniation

Postchlorination or postammoniation is the addition of chlorine or ammonia to the clearwell just prior to water being pumped to the distribution system. This is the last process used in the treatment plant operation. Chlorine and ammonia are added here to provide the residual mandated by the regulatory authorities for the distribution system and to meet $C \times T$ requirements. Although free chlorine in the clearwell and distribution system may be the residual of choice based on disinfection strength, combined chlorine is becoming increasingly popular in the operation of treatment plants to reduce the formation of DBPs and maintain a detectable residual.

Many plants are adding ammonia in conjunction with chlorine, while some add chlorine in the clearwell and ammonia at the clearwell discharge to provide a distribution system residual. A chlorine–ammonia ratio of three to four parts of chlorine to one part ammonia is used. This ratio must be maintained to keep the chloramine formed as monochloramine. Since monochloramine is the least volatile of the chloramines and has the least offensive taste and odor, any combined chlorine taste-and-odor problem is minimized. Long-term residence of chlorine in the distribution system that may be conducive to DBP formation with free chlorine is muted by the use of the combined chlorine (Figure 4-3).

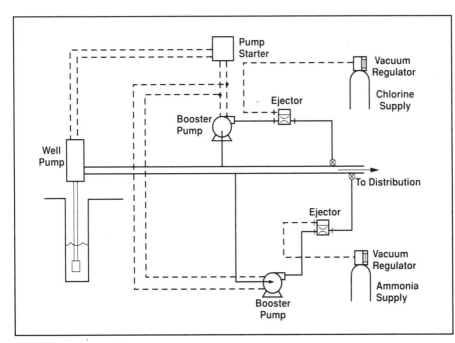

Figure 4-3
Well pump with chlorine/ammonia feed

Superchlorination

Superchlorination is the practice that calls for the addition of chlorine in dosages higher than 10 mg/L, but generally lower than 20 mg/L. Superchlorination is usually practiced in treatment plants that have large reservoirs with little changeover and a great deal of surface runoff as the source. The point of addition is generally at the raw water intake or flash mixer so that the longest contact and reaction time available is obtained. The use of superchlorination ensures that the plant water will not have the tastes and odors commonly associated with water containing high levels of ammonia. Chlorine dosages in the 5 to 7 mg/L range may create high levels of di- and trichloramine.

The practice of superchlorination has all but disappeared in the United States but is still used in Europe, particularly in Great Britain. The reduced use in the United States is driven by the desire to reduce DBPs. If the superchlorination process is practiced or planned, great care must be exercised to avoid the production of unacceptable DBP levels. Although not recognized as a typical step in drinking water treatment, superchlorination has been used on filters during startup and shutdown.

Dechlorination

Dechlorination is the practice of partially or totally removing the excess chlorine from chlorinated water. Although there may be other situations, the practice of superchlorination directly relates to the dechlorination process because superchlorination provides an excess of chlorine that must be removed.

Dechlorination is performed by using a removing agent to react with the excess oxidizing agent. The reaction between any of the forms of chlorine and the reducing agent is instantaneous and requires only good mixing to accomplish a satisfactory result. The most commonly used reducing agent is sulfur dioxide, a compressed gas available in cylinders similar to those for chlorine and ammonia. Reducing agents include sodium bisulfite and sodium sulfite solutions. Sodium metabisulfite powder, sometimes called sodium pyrosulfite or anhydrous bisulfite of soda, may also be used. The reactions of these dechlorination or reducing agents are illustrated in the following equations for each agent (Eq 4-2, sulfur dioxide; Eq 4-3, sodium bisulfite; Eq 4-4, sodium sulfite; and Eq 4-5, sodium metabisulfite):

$$SO_2 + Cl_2 + 2H_2O \rightarrow 2HCl + H_2SO_4 \qquad (4\text{-}2)$$

$$NaHSO_3 + Cl_2 + H_2O \rightarrow NaHSO_4 + 2HCl \qquad (4\text{-}3)$$

$$Na_2SO_3 + Cl_2 + H_2O \rightarrow Na_2SO_4 + 2HCl \qquad (4\text{-}4)$$

$$Na_2S_2O_5 + 2Cl_2 + 3H_2O \rightarrow 2NaHSO_4 + 4HCl \qquad (4\text{-}5)$$

The amounts of each of these reducing agents required for dechlorination are identified in Table 4-3. The values in Table 4-3 are stoichiometric quantities, and some excess capacity should be made available in the design of the system. Other materials, such as carbon (C) or sodium thiosulfate ($Na_2S_2O_5$), may be used but common practice is limited primarily to sulfur dioxide and the salt solutions. The choice of material is generally dictated by the economics of the process. There is some influence on the pH or alkalinity of the water that must be watched. Each of the reducing agents will consume some alkalinity and the process pH must be observed to ensure that pH balance is not upset.

Rechlorination

It may be necessary to add chlorine to water after it has been sent to the distribution system. This addition, performed in the distribution system, is called *rechlorination*. Long distribution systems, large finished water reservoirs, or systems that provide only a free chlorine residual can easily lose some or all of the chlorine residual in the water. Growths in the distribution system will have a chlorine demand and can remove much of the chlorine residual. This could leave the distribution system without any chlorine residual. When this happens, the utility must take corrective action by providing for additional points of chlorination at appropriate locations in the distribution system. Continuous monitoring of the chlorine residual in the distribution system or frequent sampling at specific locations in the system will alert operating personnel to a reduction in or absence of a residual. This sampling or monitoring can also be used to determine the location where the residual has been consumed. Rechlorination stations may be located at any appropriate position in the system and can be automated to operate when necessary. Control schemes for rechlorination are discussed in chapter 8.

Agent	Part Dechlorinating Agent[*]	Part Alkalinity Consumed[*]
Sulfur dioxide	0.9	2.8
Sodium bisulfite	1.46	1.38
Sodium sulfite	1.77	1.38
Sodium metabisulfite	1.34	1.38

Table 4-3
Dechlorination agent requirements

*Per part of chlorine.

Source: AWWA (1973).

Groundwater

Groundwater benefits from natural filtration. As a result, the waters pumped from wells or obtained from artesian wells and springs are often low in concentrations of synthetic contaminants. Groundwater often does have materials dissolved in it or acquires materials during its passage through the subsurface filtration system. These materials are normally inorganic, which in the presence of chlorine, do not react to form THMs. Since groundwater generally contains no organic contami- nants, THMs typically will not be present in the finished water. The inorganics in well water are due to the dissolving of the inorganic salts in the water as it filters through the underground strata. Dissolved inorganics include iron, manganese, sulfides, magnesium, calcium, and others. As mentioned previously, water from some subsurface sources can contain dissolved organics that contribute color, taste, and/or odor to the water and reacts with chlorine to form DBPs. The sources of the organics are usually swamps and forests located at the underground source. It is not unusual to provide a full-scale treatment plant for this type of water. Where some underground sources have become contaminated by synthetic materials, such as trichloroethanes (TCEs), additional treatment with advanced oxidation processes, including ozone, ultraviolet radiation, and hydrogen peroxide is often necessary.

Assuming a good quality source, chlorination is generally the only treatment employed in the treatment of well water. In these cases, chlorine is added to oxidize the unwanted inorganics (e.g., sulfide, iron, and manganese) and provide a residual disinfectant to the water as it enters the distribution system. Iron and manganese are not considered to be health risks but contribute to the formation of stains, bad taste, and iron bacteria growth in the distribution system. Most water systems use underground wells and require only chlorination. Wells influenced directly by surface water infiltration, such as those shallow wells in close proximity to rivers, may find their traditional treatment techniques altered to meet the Surface Water Treatment Rule (SWTR). The SWTR includes new filtration regulations that may require the installation of filters for this type of groundwater source. A typical well chlorination and ammoniation system is illustrated in Figure 4-3. Calculations to determine the chlorine or ammonia feed rate for the desired dosage and water flow are illustrated in Eq 4-5, 4-6, and 4-7.

$$\text{gpm} \times \text{dosage} \times 0.012 \ = \ \text{lb/d} \tag{4-6}$$

Where:

gpm = well flow, in gallons per minute
dosage = parts per million (ppm)

$$0.012 = \text{a constant}$$
$$\text{lb/d} = \text{pounds per day of chlorine or ammonia required}$$

$$\text{L/min} \times \text{dosage} \times 60{,}000 = \text{g/h} \qquad (4\text{-}7)$$

Where:

$$\text{L/min} = \text{well water flow, in litres per minute}$$
$$\text{dosage} = \text{milligrams per litre (mg/L)}$$
$$60{,}000 = \text{a constant}$$
$$\text{g/h} = \text{grams per hour of chlorine or ammonia required}$$

$$\text{m}^3/\text{h} \times \text{dosage} = \text{g/h} \qquad (4\text{-}8)$$

Where:

$$\text{m}^3/\text{h} = \text{well water flow, in cubic metres /hour}$$
$$\text{dosage} = \text{grams/cubic metre (g/m}^3)$$
$$\text{g/h} = \text{grams per hour of chlorine or ammonia required}$$

Any of these equations allows for the determination of the chlorine or ammonia feed rate, depending on the water flow and the dosage desired.

In most well water systems, water is maintained under pressure from the well and may pass through a pressure filter system to remove oxidized inorganics before entering the distribution system. Since the chlorine demand in the water is often constant, the chlorination system need only be set manually for the correct dosage and activated automatically with the operation of the well pump. Sufficient CT and dosage must be provided to meet the required $C \times T$ value prior to the first service connection and to meet state regulations regarding the desired residual in the distribution system.

Purchased Water

Many water utilities buy water from neighboring utilities through common or interconnected pipelines. This situation requires a special review of the residual desired by the purchaser and that available from the supplier. If both systems want a free chlorine residual, the addition of chlorine, appropriately controlled, to reach a desired residual is relatively simple. Complexities develop when waters produced by the two systems do not have the same type of residual (free or combined) or level of residual and the two waters are to be blended. The resultant mixture must be monitored for the desired residual (free or total) and the correct calculation performed to determine the amount of each chemical to be added to reach the desired residual. The determination can become more complicated with the mixing of unequal quantities of waters with different levels of free or combined chlorine. This latter situation is becoming more common today with the increased use of chloramines in the treatment process. Appendix A contains some typical calculations that must be performed to ensure the correct chemical addition. Monitors should be used to measure and confirm the result.

System Preparation

Before any water treatment system can be put into service, the system must be disinfected and prepared to receive the water. The American Water Works Association (AWWA) has developed a series of standards covering disinfection from the well to the treatment plant and from storage facilities to the distribution system. Updated regularly, these standards present accepted procedures and practices to be taken prior to placing the system, tank, piping system, etc., into service. Some of these standards are listed below.

AWWA C651, Standard for Disinfecting Water Mains

AWWA C652, Standard for Disinfection of Water-Storage Facilities

AWWA C653, Standard for Disinfection of Water Treatment Plants

AWWA C654, Standard for Disinfection of Wells

Some of the more important points taken from these standards or from the author's personal experiences include the following:

- The initial disinfection efforts should be directed toward cleaning the facilities. Pipes should be stored to prevent contamination prior to use and protected after installation. This is important since the pipe is not easily inspected once installed. Tanks, wells, and plant facilities should be thoroughly cleaned by washing with sufficient water to remove debris, dirt, etc. Protect equipment after installation to reduce the possibility of contamination.
- Regardless of the facilities to be disinfected, the type of disinfectant used is usually chlorine gas, liquid sodium hypochlorite, or calcium hypochlorite tablets. The injection of either gas chlorine or sodium hypochlorite into the flowing water provides a more uniform-strength solution.
- Surface washing of tanks and other plant facilities with solutions of 200 ppm (mg/L) of hypochlorite is frequently practiced. Whether surface washing is practiced or the walls are simply flushed with high volumes of water, the facility must be exposed to a minimum concentration of chlorine residual for a fixed time period.
- Water mains require a 24-h exposure with a minimum of 25 ppm (mg/L) of free chlorine, while tanks can be exposed to concentrations of 10 ppm (mg/L) of free chlorine with a

concentration of 2 ppm (mg/L) at the end of 24 h. Treatment plant disinfection requirements are similar to the water main and storage tank levels, while wells require a 50-ppm (mg/L) free chlorine exposure for 12 to 24 h.

- Disposal of the highly chlorinated waters at the end of the disinfection period must follow common sense and meet the appropriate regulations governing effluent discharge. The use of dechlorination agents, discharge into the local sewer system, or dilution prior to discharge may be satisfactory methods of disposal.

- Most state departments of health will provide the guidelines for exposure time and minimum concentrations to be met. The regulations of the state in which the utility is located must be known and followed.

- The residuals to be measured in these disinfection procedures are considerably higher than the operating levels at the treatment plant. Manual sampling of these levels and use of the drop dilution method to determine the residual level is necessary. The drop dilution method is covered in chapter 5.

Engineering Considerations

There are many situations in a water treatment plant that require special evaluation. Probably the most important is the intimate mixing of chlorine and ammonia solutions on addition to the water. If either chemical is not distributed evenly and equally to the water, the desired result will not be attained. The disinfection and oxidation process is only as good as the ability to achieve adequate mixing.

All plants must choose the point of addition that provides the greatest opportunity to effectively and efficiently mix the materials for their intended purpose. The chemicals to be added must be reviewed so that their interaction is known. Some chemicals added at the same location can have a counterproductive effect. This is true if chlorine and powdered activated carbon (PAC) are added at the same location. Short of using an external device (e.g., a static mixer or propeller mixer), mixing is best achieved by choosing the point where flows are the most turbulent. This allows the natural mixing action of the flow to be maximized. Addition ahead of points of direction change, pipe fittings, hydraulic breaks, pipe restrictions, and similar changes in flow patterns are also recommended.

Of all the treatment locations, injection into a contact tank presents the greatest difficulty. Contact tanks always require special consideration. The use of baffles and recirculating devices, the high length-to-width ratio and high velocity attained in narrow channels, and the use of external mixing devices greatly aid in the disinfection process. Diffusers

Figure 4-4

Gas induction
systems

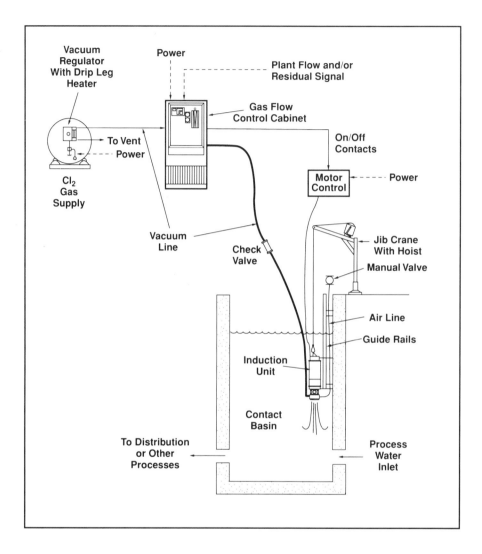

that inject across the entire stream flow and have outlets that direct the solution discharge against the flow will achieve better results.

Recently developed mixing devices that provide the needed vacuum for operation of the chlorine gas feeding devices have been used quite successfully. These devices provide the capability of producing vacuum levels in the range of 25 to 29 in. (635 to 737 mm) of mercury and feeding up to 10,000 lb of chlorine/day (200 kg/h) or 5,000 lb of ammonia/day (100 kg/h). These devices can also be employed to feed solutions of hypochlorite or ammonia compounds (Figure 4-4).

Sampling points for feed to a chlorine residual analyzer should be from a point in the water flow that provides for a uniform mixture. The location of the sampling point in a pressure pipe system should be after the mixing and reaction is complete. Sufficient sample water should be moved to the analyzer to minimize the lag time in the system from the injection point to the analyzer. Piping size and water flow quantity must be considered. If control rather than monitoring is of concern, the entire process loop must be reviewed to minimize lag time.

When a contact tank is used, common in many surface water plants, the location of the sampling point must account for different water flow rates, channeling, short-circuiting, etc. Several sampling locations may be available for use. The sampling point can be selected by the operator to accommodate the particular flow, or selected through an automated process to account for various flow rates and contact times.

References

American Water Works Association. 1973. *Manual M20, Water Chlorination Principles and Practices*. Denver, Colo.: AWWA.

Kirmeyer, G.J., G.W. Foust, G.L. Pierson, J.J. Simmier, and M.W. LeChevallier. 1993. *Optimizing Chloramine Treatment*. Denver, Colo.: AWWA Research Foundation and AWWA.

US Environmental Protection Agency. 1982. *Guidance Manual for Compliance With Filtration and Disinfection*. Prepared by Malcolm Pirnie and HDR Engineering. Washington, D.C.: USEPA.

Chlorine Residual Analyses and Measurement

All the processes dealing with chlorination and chloramination of water are of little value if the results of the effort cannot be measured. These results include not only the level of disinfection achieved and amount of inorganic and organic contaminants removed but also the ability to measure the amount of chlorine and chloramine residual remaining in the water. It's also important to be able to measure the chlorine and chloramine residuals in a simple, efficient, and rapid fashion. Since the inception of chlorination and chloramination, a great deal of effort has been expended over the years to achieve this goal. This chapter examines the sampling procedures and the analytical methods available for use in the laboratory and field and compares their advantages and disadvantages.

Definitions

We have already seen that there are many chlorine and chloramine reactions in water. In all these, the results of these reactions and the form or forms of chlorine remaining are important. In analyzing chlorine and cloramine reactions, it is useful to understand the following definitions:

Dosage is the amount of chlorine or chloramine added to water, expressed in mg/L (ppm) as Cl_2.

Demand is the amount of chlorine or chloramine compound immediately consumed by or reacted with the water's organic and inorganic components. It is expressed in mg/L (ppm) as Cl_2.

Residual is the amount of chlorine or chloramine remaining in the water, expressed in mg/L (ppm) as Cl_2.

The above concepts are expressed in Eq 5-1.

$$\text{Dosage} - \text{Demand} = \text{Residual} \qquad (5\text{-}1)$$

Chlorine residual, whether free, combined, or total, is an important factor and must be closely followed by the operator to measure the performance of the disinfection process. A chlorine residual value is, after all, the mandatory information requirement that the state regulatory agencies require from all water utilities. Although each state's reporting requirements may be different, and the form of residual (free or combined) desired may vary, the states are unanimous in requiring a reportable chlorine residual. The reason is fairly simple—a chlorine residual is a major determinant of the quality of drinking water.

Laboratory Methods

Larger water utilities will have available well-equipped laboratories to perform the routine and not-so-routine water analyses, including chlorine residual measurement. In general, the larger a water utility, the more sophisticated its capabilities. Conversely, the smaller the utility, the more basic and simple its capabilities. Small utilities may contract their analysis needs to outside laboratories. If this is done, it is necessary to ensure that the laboratory is certified and approved for practice in the respective states.

The method used for chlorine residual measurement can be simple or sophisticated depending on the analytical requirement or reporting needs. The US Environmental Protection Agency (USEPA) and the states have established reporting requirements based on the number of people served and the source of the water, i.e., surface or ground. If your state has assumed primacy, it is important to know what the state regulations require. This is usually accomplished by contacting your appropriate regulatory authority. If your state has not assumed primacy, then the responsible USEPA office must be contacted to confirm the applicable regulations and the reporting requirements. The 1986 Surface Water Treatment Rule (SWTR) and its subsequent amendments dictate the disinfection by-product requirements. These rules were under discussion at the time of this writing.

Standard Methods for the Examination of Water and Wastewater (APHA, AWWA, and WEF 1995), commonly referred to as *Standard Methods*, is the source of the approved analytical methods of water analysis to determine free, combined, and total chlorine residuals. This book has been published since 1905 under the auspices of the American Public Health Association (APHA), American Water Works Association (AWWA), and the Water Environment Federation (WEF). APHA initiated the publication and was joined by AWWA in 1933 and WEF (in the form of its predecessor, the Water Pollution Control Federation [WPCF]) in 1955. All laboratories and water utilities should consider having this book as a standard part of their libraries. Updated issues are published on a periodic basis to provide the latest, approved procedures. USEPA has also been developing some testing methods, but these are not seen as widely divergent from the methods presented in *Standard Methods*.

Many procedures relating to residual analysis have been proposed and used over the years. Some have been added or dropped due to improved technology or more rigid requirements. All procedures have controllable parameters, such as temperature, time, and interferences, that must be followed. Complex chemistry practices require skilled, trained operators. Each procedure lists these limitations, and the individual performing the analyses should be aware of them.

The current methods of chlorine residual analyses that seem to be the most frequently used are the amperometric and colorimetric methods. Both require equipment and analytical skill to perform. The equipment needed may require the investment of a considerable amount of money. The skill required must be developed over time with practice and patience. Once developed, the skill must be maintained at a high level of precision by frequent practice. Each step in the analysis procedure must be followed exactly as the method prescribes.

The currently approved laboratory methods are the iodometric (methods I and II), amperometric titration (Figure 5-1), low-level amperometric, DPD (*N,N*- diethyl-*p*-phenylenediamine) ferrous titrimetric, DPD colorimetric, syringaldazine (FACTS) [free available chlorine test, syringaldazine]), and iodometric electrode. For a detailed discussion of the procedures, equipment, and chemicals required by these methods, refer to *Standard Methods*, Part 4500. The following provides a summary of the various laboratory methods:

Iodometric (methods I and II)

Type measured	Total chlorine
Technique	Titration with sodium thiosulfate
Interferences	Manganese oxide, other oxidizing agents, reducing agents, organic sulfides, ferric and nitrite ions

Figure 5-1
Two null balance amperometric titrators. The titrators can be used to identify the free, combined, and total species.

Source: Bailey-Fischer and Porter.   *Source: Wallace & Tiernan Inc.*

Normal range	1 mg/L using the starch iodide endpoint, 40 µg/L using 0.01 normal thiosulfate
Comments	Relatively simple to perform

Amperometric

Type measured	Free, combined, or total chlorine, mono- and dichloramine
Technique	Titration with phenylarsine oxide (PAO) in conjunction with an amperometric detection cell

Interferences	Other halogens, chlorine dioxide, high-speed mixing, nitrogen trichloride
Normal range	0.1 to 2 mg/L. Higher ranges use a diluted sample
Comments	Requires a higher degree of training than iodometric method

Low-level Amperometric

Type measured	Total chlorine
Technique	Titration with PAO
Interferences	Same as amperometric
Normal range	Less than 100 down to 10 µg/L
Comments	Usually not used in water treatment applications but rather in wastewater

DPD Ferrous Titrimetric

Type measured	Free, combined, or total
Technique	Titration with ferrous ammonium sulfate
Interferences	Strong oxidants, manganese oxide, copper, chromate
Normal range	More than 0.1 mg/L
Comments	Somewhat difficult to perform, requires mathematical calculations to determine different fractions

DPD Colorimetric

Type measured	Free, combined, or total
Technique	Colorimetric comparator, spectrophotometer
Interferences	Strong oxidants, color, turbidity, chromate
Normal range	More than 0.1 mg/L
Comments	Comparator is very subjective and not as precise or accurate as titration methods

Syringaldazine (FACTS)

Type measured	Free
Technique	Colorimetric
Interferences	Color, turbidity, other oxidants
Normal range	0.1 to 10 mg/L
Comments	Very subjective and not as precise or accurate as titration methods

Iodometric Electrode

Type measured	Total
Technique	Titration
Interferences	Oxidizing agents, manganese oxide, iodate, bromine, cupric ions
Normal range	1 mg/L and up
Comments	Not direct reading

According to an AWWA survey, the DPD method is the most frequently used laboratory method. The capability of this method to determine free and combined chlorine quickly and economically makes the choice relatively easy. For accuracy, the amperometric method is the most popular with laboratory personnel. Commercially available titrators are available from several sources and the chemicals required are standard at most laboratory supply houses.

Field Methods

The conditions that water department personnel find themselves confronting in the field, away from the friendly confines of the analytical laboratory, dictate a different set of operating parameters. The lack of power and running water calls for equipment that is portable, convenient, easy to handle, and quick. The loss of accuracy that tends to accompany these criteria is offset by the needs of rapid, easy, field reading. Field-determined results do not provide the accuracy of laboratory testing.

As a result, field testing is most often done using the colorimetric testing procedures. There are many commercially available colorimetric test kits that are reliable across a wide range of residuals. The selection of the range is up to the user; the manufacturer should be consulted as to styles and ranges available. Ranges up to 2 or 3 mg/L are recommended for water systems. Each range has anywhere from 5 to 10 readings possible. There may be a need for low-range readings, particularly when attempting to determine the free and combined residual portions.

Some years ago, all colorimetric kits used were based on the orthotolidine (OT) test method. This test has been discontinued due to the poor ability of orthotolidine to distinguish between the combined chlorines and free chlorine. The combined chlorines react slowly with OT and do not provide the quick, clearly defined result required. There also is a concern regarding the suspected carcinogenicity of OT.

Current methods use colors generated by reaction with the DPD. These colors are clear, develop rapidly, and provide the ability to determine free, mono-, di-, and trichloramine values that were previously unavailable with colorimetric test kits. The kits are relatively inexpensive, compact, and easily transportable. Field distribution personnel will usually carry these kits with them at all times.

The colorimetric kit method does have some interferences, including turbidity and color. In addition, since the reported values depend on human interpretation, the readings are subjective and not as accurate. Battery-operated colorimeters can provide objective values and improve the accuracy. Normally readings are taken to the nearest 0.1 mg/L, depending on the kit and the calibrated color comparisons provided. A blank must be run to aid in zeroing out the interferences caused by the water being analyzed. Procedures for the tests are provided by the manufacturers and must be followed consistently and exactly to ensure accuracy. Whatever pains are taken to follow the procedures, the use of a field test kit produces results less accurate than laboratory analyses. If accurate results are desired, samples should be returned to the laboratory as expeditiously as possible for evaluation.

Drop Dilution Method

As identified in chapter 4, the sterilization of mains, storage tanks, and the water treatment plant itself requires high concentrations of chlorine solution in the range of 50 mg/L or more. The measurement of these values requires some modifications to the standard analytical techniques. This modification is called the *drop dilution method*.

The drop dilution method is used for concentrations higher than 10 mg/L when a standard colorimetric test kit is used (although there are some kits available for these ranges). The method requires the use of demand-free water that dilutes the sample by a predetermined amount and reduces the sample's value to the range covered by the kit. The procedure is described in AWWA C651-92 (AWWA 1992b) and *Standard Methods*. The test kit reading is adjusted by the following calculation:

$$\text{true residual} = \frac{\text{reading} \times 200}{\text{sample drops}} \qquad (5\text{-}2)$$

67

Where:

true residual mark = calculated residual, in mg/L
sample drops = the number of drops of water from disinfected water added to the test tube or vial
200 = a constant to account for the volume of each drop
reading = the test kit reading, in mg/L

Chlorine Demand Test

The calculation of the amount of chlorine required to treat water can be determined by the chlorine demand test. This test is covered in *Standard Methods*, Part 2350B. The chlorine demand of water to be treated can sometimes be predicted by analyzing the constituents in the water and calculating the chlorine feed rate. This is easily done if the constituents are primarily dissolved materials, such as iron, manganese, and sulfides. The exact quantities required are known from the chemistry or stoichiometry of the chemical reaction. However, if constituents include organics, as in surface waters, it may not be possible to predict the dosage, and the chlorine demand of the water must determined. The determination of the chlorine demand requires the use of either the amperometric titration or the DPD test kits. The choice of the test depends on the accuracy needed. In general, if the test is to be run, the amperometric titration is recommended. Because of the concern over THMs, further analysis to determine the THM formation potential may also be required at the time the demand test is done.

Sampling

One of the most critical items for any residual testing is the sampling method used. Proper sampling techniques are identified in *Standard Methods*, Part 1060. Primary considerations of the procedure include the following:

Sample location	Choose the location that will provide the most representative sample.
Source consistency	Flush the sample line to ensure that the sample is representative of the source.
Sample container	Ensure that the sample container is flushed clean with the sample water and filled to the top, leaving no air space.
Sample time	Ensure that the sample is tested as soon as possible after taking the sample.

| Sample protection | Protect the sample from light and heat, cap until tested, and store at a low temperature. |

Continuous Analysis

Equipment is available for the continuous measurement of chlorine residual in water. The equipment used is generally based on the amperometric or colorimetric techniques. Details on this type of equipment can be found in chapter 7.

References

American Public Health Association, American Water Works Association, and Water Environment Federation. 1995. *Standard Methods for the Examination of Water and Wastewater*. 19th ed. Washington, D.C.: APHA.

American Water Works Association. 1992a. Survey of Water Utility Disinfection Practices. *Jour. AWWA*, 84(a):121–128.

———. 1992b. *AWWA C651-92, Standard for Disinfecting Water Mains*. Denver, Colo.: AWWA.

Chemical Handling and Safety

This chapter deals with shipping, handling, storage, and use of chlorine, ammonia, the hypochlorite solutions, aqueous ammonia, and ammonia salt solutions. Regulations governing these operations and recent developments in fire codes are also discussed.

Chlorine

The US Department of Transportation (DOT) regulates the shipping and handling of chlorine on land, while the US Coast Guard regulates shipping and handling on waterways. Shipping containers also come under the jurisdiction of DOT, which classifies chlorine as a poisonous gas. Although state and local regulations and insurance requirements should also be reviewed, federal regulations usually prevail. An example is in the shipment and handling of containers containing chemicals for which DOT is responsible, not state or local fire codes. However, it is prudent to know local regulations and comply with them in cases where no conflict exists with federal regulations.

The great majority of water treatment facilities use chlorine in 150-lb (68-kg) cylinders or ton containers. Larger consumers use railroad tank cars or employ over-the-road tank trucks. The tank trucks either unload directly into the process or into on-site storage tanks. The delivery/storage method used is a function of the quantity of the chemicals used, the cost of the chemicals, and the ability of the facility to handle either the containers or bulk delivery.

Since most chlorine installations consume less than 100 lb of chlorine a day (2 kg/h), 150-lb (68-kg) cylinders are the most frequently used containers. For example, a 2-mgd (991-m^3/h) well treated with 2 ppm (mg/L) of chlorine would only consume about 32 lb of chlorine a day (640 g/h) in a continuous operation. This means that the well would need the replacement of a 150-lb (68-kg) cylinder about every 5 days. This assumes that the well runs 24 h a day, 7 days a week—not a likely possibility. A 5-mgd (2,477-m^3/h) surface water treatment plant with a 2-ppm (mg/L) dosage, would require approximately 84 lb of chlorine a day (1.68 kg/h). This, too, is well within the capability of 150-lb (68-kg) cylinders.

An advantage of ton containers is the reduced cost of chemicals. This advantage must be balanced against increased handling needs, possibly larger storage areas, the safety implications of ton storage, and any new design conditions introduced.

Chlorine Cylinders

Chlorine cylinders are seamless steel containers designed according to DOT specification 3A480 or 3AA480.[*] Cylinders contain a maximum of 150 lb (68 kg) of chlorine, although smaller sizes can be obtained. Each cylinder is equipped with a valve approved by The Chlorine Institute. By law, only 80 percent of the cylinder volume is used when the cylinder is filled. The unused 20 percent is to allow for the expansion of liquid chlorine due to any temperature increase. With chlorine's high coefficient of expansion, proper measures must be taken to ensure that the expansion of the liquid in the cylinder does not cause a hydrostatic rupture of the container. Gas is withdrawn when the cylinder is standing on its base. Although not a common practice, liquid may be withdrawn with the cylinder inverted (Figure 6-1).

Each cylinder uses a cylinder valve that contains an insert called a fusible plug. This plug contains a special metal alloy, a eutectic, that will melt when the chlorine temperature reaches between 158 and 165°F (70 and 73.9°C). At this temperature, the chlorine in a cylinder filled to the 80 percent mark would expand to fill the vessel to 100 percent of its volume (Figure 6-2). The melting of the fusible plug releases the chlorine to relieve any pressure that would exceed 300 psig (2,068.5 kPa [gauge]). The chlorine vapor pressure curve presented on page 11 (Figure 2-1) plots chlorine pressure versus temperature. The chlorine pressure is

[*]Title 49, Code of Federal Regulations. Available from Superintendent of Documents, US Governmental Printing Office, Washington, DC 20402.

300 psig (2,068.5 kPa [gauge]) at the melting point of the fusible plug. If there was no fusible plug, then the cylinder would be exposed to higher pressures and could rupture, thus creating the potential for far greater damage.

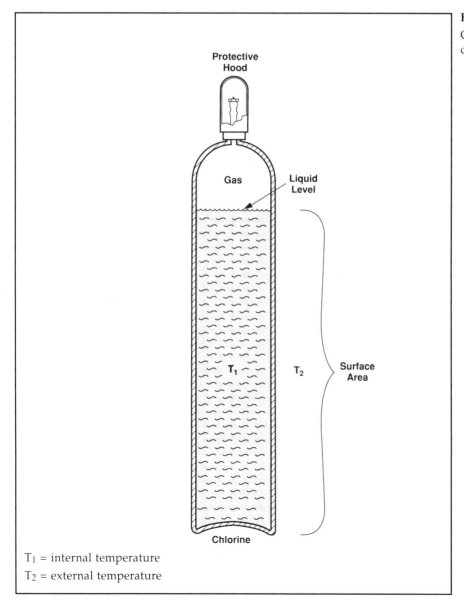

Figure 6-1
Chlorine cylinder

T_1 = internal temperature
T_2 = external temperature

Ton Containers

Ton containers hold, as the name implies, 2,000 lb (907 kg) of chlorine. They are shipped and used in the horizontal position. At one end of the container are two valves. When these valves are positioned so that a line drawn between them is perpendicular to the horizontal, the top valve is used for chlorine gas feed and the bottom valve is used for chlorine liquid feed. The valves on ton containers are different than those used on cylinders. Ton valves have a larger opening than most cylinder valves to allow for larger feed rates. Ton valves do not contain the fusible plugs used on cylinders. Instead, there are three fusible plugs located at

Figure 6-2

Volume–temperature relation of liquid chlorine in a container loaded to its authorized limit

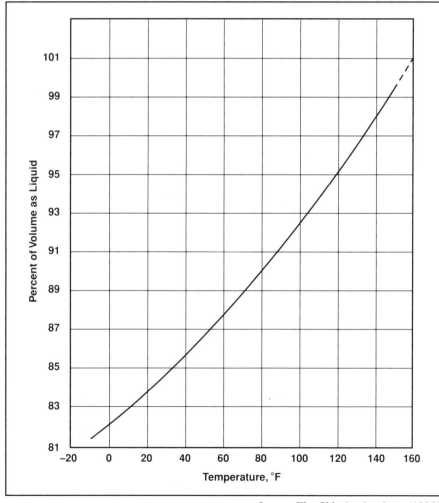

Source: The Chlorine Institute (1996).

each end of the ton container. The plugs are positioned at 120° angles to each other toward the outer edge of the container. In addition, each valve is connected to a tube, commonly called an *eduction tube*, that reaches to the container wall to ensure that gas or liquid is always available to the valve even as the container empties. As with the 150-lb (68-kg) cylinders, ton containers are initially filled only to 80 percent of the available volume (Figure 6-3).

Tank Cars

Tank cars usually have capacities of 55 to 90 tons (49,885 to 81,630 kg), but additional sizes also may be available depending on location or need. All tank cars are equipped with 4-in. (101-mm) insulation surrounding an inner steel tank. The cars are equipped with two liquid discharge valves and two valves connected to the vapor phase. Each car is equipped with a safety relief valve set to open at either 225 or 375 psig (1,551 or 2,585 kPa [gauge]) depending on the tank car. The liquid discharge lines contain excess-flow valves ahead of each liquid discharge valve to stop the flow of chlorine and prevent any further discharge in the event of a downstream line break. The excess flow valves are designed to seat at flows of 7,000, 10,000, and 15,000 lb/h (3,178, 4,540, and 6,810 kg/h). Although gas discharge lines are provided, they are not used to unload chlorine. The gas discharge lines are used primarily to enable the tank car to be padded to a higher pressure to aid the removal of liquid chlorine. Gas removal through the gas lines is limited due to the insulation on the car. This insulation greatly reduces the external heat reaching the liquid chlorine and lowers the rate of vaporization of the gas in the tank car (Figure 6-4).

According to DOT regulations, tank cars must always be attended during unloading and no more than one tank car may be connected to an unloading facility at the same time. Any deviation must receive an exemption from DOT. Since tank cars are supported on the railcar wheels by heavy-duty springs, connections to tank cars must be made of flexible material to accommodate tank cars rising due to the loss in weight.

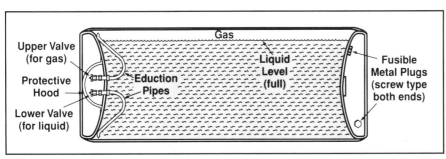

Figure 6-3
Ton containers

Labels:
Upper Valve (for gas)
Protective Hood
Lower Valve (for liquid)
Gas
Liquid Level (full)
Eduction Pipes
Fusible Metal Plugs (screw type both ends)

Gas Withdrawal

Withdrawal of gaseous chlorine from either cylinders or ton containers must be handled differently than withdrawal of liquid chlorine. Gaseous chlorine is removed by using the pressure in the cylinder as the driving force to move the gas. This requires the availability of sufficient heat to vaporize the liquid chlorine and maintain the gas phase, regardless of the type of container. As gas removal proceeds, the liquid temperature decreases and the pressure in the

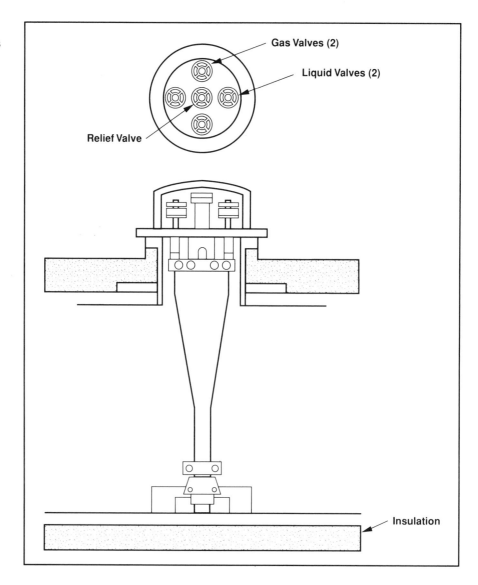

Figure 6-4

Tank car cross section, valve arrangement, and manway

container decreases. A drop in container temperature and pressure can cause a decrease in the chlorine withdrawal rate. Condensation appearing on the surface of the container indicates that the liquid temperature in the container has reached the dew point of the surrounding air. This may be cause for concern because the desired feed rates may not be maintainable. As gas withdrawal continues, the liquid temperature in the container continues to decrease. When the liquid temperature reaches 32°F (0°C), the condensed moisture on the container's exterior walls will freeze and form ice. The formation of ice is to be avoided because the ice will act as an insulator and further impede heat transfer into the container. The chlorine pressure and temperature will continue to decrease and the chlorine feed rate will decrease rapidly and eventually cease. A curve of temperature versus gas feed rates can be used to estimate gas withdrawal capability (Figure 6-5). Although the Chlorine Institute (1986) advises that the "dependable, continuous discharge rate from a cylinder" is about 1.75 lb/h (0.8 kg/h) or 42 lb/d (1.4 kg/h), experience shows higher feed rates are possible. The Chlorine Institute assumes an ambient temperature of 70°F (21.1°C) for this feed rate and the pressure against which the gas is being discharged is 35 psig (241.3 kPa [gauge]). According to the Institute, these feed rates can be

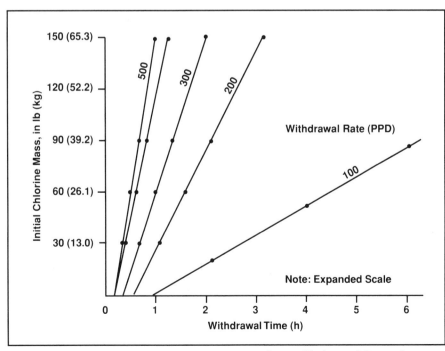

Figure 6-5
Cylinder gas withdrawal rates, withdrawal curve at 80°F (26°C) ambient temperature

Source: Fischer and Porter Company.

doubled if air circulation is provided around the cylinder. This minimizes condensation, aids heat transfer, and, thus, improves feed rates.

With today's direct cylinder-mounted chlorinators acting as vacuum regulators, the restriction to gas flow is reduced further and gas feed rates of 100 lb/d (2 kg/h) are attainable. The Chlorine Institute states that the feed rate from ton containers is 15 lb/h (6.8 kg/h) or 360 lb/d (163.2 kg/h), which can be doubled if air circulation is provided. In addition, the feed rate is enhanced with direct ton-container-mounted chlorinators, which reduce the backpressure on discharge from the container to a vacuum at the point of exit. It is possible to achieve 500 lb/d (10 kg/h) feed rate.

Liquid Withdrawal

Liquid withdrawal usually is unaffected by ambient temperature as long as the container pressure provides sufficient pressure differential to move chlorine to the receiving point. Liquid withdrawal requires the use of chlorine vaporizers since, in the water industry, chlorine is fed through chlorine gas feeders. Chlorine gas feeders must receive only gas; they will malfunction if liquid chlorine is permitted to reach the feeder.

Physiological Effects of Exposure to Chlorine

Chlorine gas can be sensed in the respiratory system, eyes, skin, mouth, and mucous membranes. In the presence of moisture, acids will form on the skin surface and irritate the skin. The skin will redden and may blister, depending on the severity of the irritation. Liquid chlorine exposure is far more severe and can cause thermal burns and severe blistering. Chlorine's impact is a function of exposure time and concentration. The physiological effects from exposure to various levels of gaseous chlorine taken from information and assessments by the Occupational Safety and Health Administration (OSHA) and the American Council of Governmental Industrial Hygienists (ACGIH) are presented in Table 6-1.

Chlorine Piping

Chlorine liquid or gas pressure piping is normally 0.75-in. (18-mm) or 1-in. (25-mm) diameter carbon steel, seamless schedule 80 material meeting ASTM A106 grade B specifications (ASTM 1993). Chlorine gas under vacuum or up to 6 psig (463 kPa [gauge]) allows for the use of plastic pipe or tubing constructed of polyvinyl chloride (PVC), chlorinated polyvinyl chloride (CPVC), acrylonitrile-butadiene-styrene (ABS), polyethylene (PE), polypropylene (PP) or fiberglass-reinforced polyester (FRP) materials. For chlorine solution (the mixture of chlorine and water from the chlorine ejector/venturi), PVC, CPVC, ABS pipe, ABS

Exposure Level (ppm)	Effects	Table 6-1
0.2–0.3	Odor detectable by most people	Physiological effects of chlorine
< 0.5	No known acute or chronic effect	
0.5	8-h time-weighted average (TWA) (ACGIH*)	
1.0	OSHA ceiling level short-term exposure level (STEL)	
1–10	Definite odor detectable, slight irritation of eyes	
10	Immediately dangerous to life or health (IDLH) (NIOSH†)	
15	Immediate irritation of nose, throat, and eyes, accompanied by coughing	
100	Potentially lethal (depending on duration of exposure)	
1,000	Dangerous to life after a few inhalations	

Source: The Chlorine Institute (1995).

*American Conference of Governmental Industrial Hygienists.
†National Institute of Occupational Safety and Health.

hose, rubber hose, or PE hose suitable for the pressure intended can be used. *Due to the corrosive nature of chlorine solutions, steel or iron pipe must not be used to carry chlorine solutions unless a suitable corrosion-resistant lining is employed.*

Chlorine cylinders or ton containers can be connected to a common manifold for gas withdrawal. Each container should have an isolating valve connection to permit independent servicing of each container while the system is in operation. Liquid withdrawal from manifolded containers is treated differently and discussed later. For further details related to chlorine piping needs, the reader is referred to *Piping Systems for Dry Chlorine* (The Chlorine Institute 1992).

Reliquefaction

Reliquefaction is the condensation of chlorine gas on the internal walls of the gas pressure piping. Of all the conditions to be avoided in chlorine pressure piping installations, gas condensation causes the most concern. As the gas condenses, droplets of liquid chlorine form on the piping surfaces. These droplets can be carried along with the gas stream to the gas feeder. When liquid droplets reach the gas feeder, they will attack the gas feeder's plastic materials. Prolonged exposure will soon make the feeder inoperative. Reliquefaction must be avoided.

All chlorine gas pressure piping must be installed in such a way as to prevent reliquefaction. Proper protection can include the use of heat tracing and insulation from the container to the feeder. The use of pressure-reducing valves to lower the gas pressure, which lowers the liquefaction temperature, is also helpful. Figure 6-6 illustrates a recommended method of installation. Additional protective methods include the use of drip legs in the gas line at points of direction change and adding a low-wattage heater on any drip legs. Gas pressure piping should always be sloped toward the source to aid droplet drainage back toward the containers.

Installations should be reviewed to ensure that the pressure line will pass through increasing temperature zones as the gas moves from the source toward the feeder (Figure 6-6). If lower temperature zones are reached, the chance of reliquefaction is increased. Routing gas pressure pipe near windows, through basement or underground areas, or anywhere cooler temperatures may be encountered, should be avoided.

When containers are connected to pressure piping, the connection is made with a flexible connector. Any connection to the container valve must be made with a yoke and adaptor connection of either the open or

Figure 6-6

Gas pressure piping system

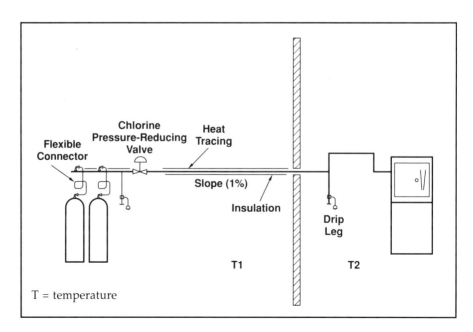

T = temperature

closed type (Figure 6-7). The use of threaded (CGA [Compressed Gas Association] 660) connections for container hookup should only be with permanent or semipermanent connections.

Regardless of the connection, a new gasket is required with each change. Although fiber-type gaskets are available, lead-type gaskets are recommended because the gaskets do not recover after compression and are not reusable. New lead gaskets must be used with each change. This provides some measure of security in chlorine installations and forces the operator to change the gasket. Since fiber gaskets can recover from compression, this type of gasket may be reused. This reuse frequently causes fraying at the edges. Loosened fibers my be carried by the gas stream into the gas feeder where they will clog the orifices of the feeder and can affect gas feed. Since lead gaskets do not act in this fashion, they are more desirable for use with gas feeders.

Flexible connectors may be connected and disconnected repeatedly as long as good practices and common sense are used. The connectors should not be kinked as this will create weak spots in the connector. Flexible connectors, sometimes referred to as pigtails, are made from copper tubing and may be covered with corrosion protection, such as cadmium plating. The connector should be examined after each use to observe the appearance of any greenish buildup, which indicates corrosion. When changing container connections, the movement of the connector may produce a creaking noise. This sound indicates corrosion and suggests it is time to consider replacing the connector. The more frequently containers are changed, the more frequently the connectors should be changed. At the very least, the connectors should be changed annually. There are also other types of connectors available for use on gas pressure pipes. Refer to The Chlorine Institute (1992a).

There is a distinct advantage to vacuum piping when using gas feeders. The safety of the system is improved because the pressure

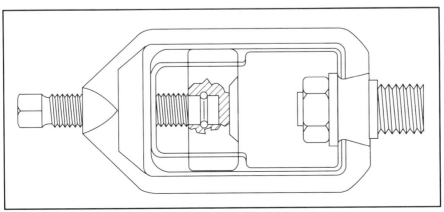

Figure 6-7
Yoke assembly
with nipple

portion of the system is limited to one connection at the container valve. Any break in the vacuum line or other loss of vacuum will stop the flow of chlorine. If any break occurs in a pressure line, the flow of chlorine gas will continue creating an undesirable condition. The plastics used in vacuum-line installations are generally more economical to install and provide a resistance to corrosion if moisture enters the system. In addition, since the reliquefaction temperature is considerably reduced, the chance of liquid droplets reaching the gas feeder has been minimized. The reliquefication point of chlorine is about $-30°F$ ($-34.4°C$) at atmospheric pressure.

Liquid Chlorine Piping Systems

Liquid chlorine piping systems require some different considerations than those of gaseous chlorine systems. One important concern is protection against liquid chlorine expansion. When liquid chlorine can be isolated or trapped between two closed valves, there is danger of hydrostatic rupture due to the high coefficient of expansion of chlorine. All such situations call for protection using a chlorine expansion chamber. Figure 6-8 illustrates one type of expansion chamber, complete with pressure gauge and pressure switch. When installed in a liquid line, both a visible indication and a contact closure for remote alarm/indication are provided. The chamber must be sized so that it contains 20 percent of the volume of the section of the liquid line it is protecting.

In addition to expansion chambers, there are occasions when installations may need to withdraw liquid from ton containers for feed to vaporizers. Multiple-container, liquid hookup may be desired. To accommodate such an installation, gas pressure equalization between the containers feeding the manifolded liquid line is required. Figure 6-9 illustrates the correct method of piping when ton containers' liquid valves must be manifolded. The manifold on the gas valves is not intended to feed gas but to provide a pressure equalization line between the containers connected to the liquid manifold. If the pressures are not equalized, unequal distribution of liquid chlorine between the containers is possible. Should one of the containers be overfilled and the container valves shut, the container could suffer a hydrostatic rupture. The use of the gas manifold requires a thorough review of the installation to ensure that the Chlorine Institute recommendations are understood and applied where appropriate (refer to Drawing 183, The Chlorine Institute 1986).

All instrumentation on both liquid chlorine and gaseous chlorine lines should have suitable protection. Gauges and switches must be provided with isolating diaphragms. The nonprocess side of the instrument should be protected with a liquid-filled seal containing a fluorocarbon that is not reactive with chlorine. The diaphragms should be constructed of corrosion-resistant materials, such as tantalum,

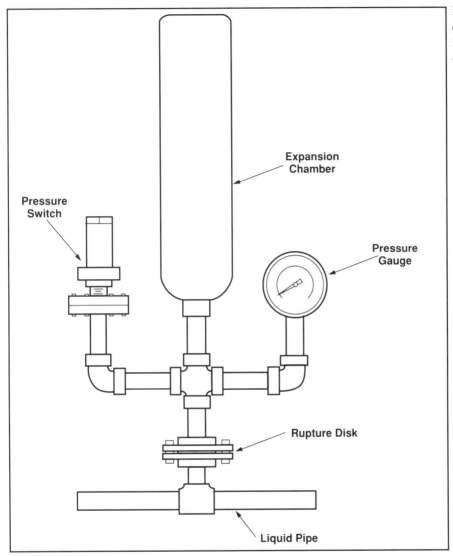

Figure 6-8
Chlorine liquid
line expansion
chamber

hastalloy C, or monel. Instrumentation on chlorine solution lines from the chlorinator ejector/venturi must also be protected from corrosion by the acid–chlorine solution. The pH of the solution line will be in the range of 2 to 3 and housings of PVC or a suitable corrosion-resistant material containing diaphragms of Viton® or a similar corrosion-resistant material will be necessary.

Figure 6-9

Container liquid
manifold
arrangement

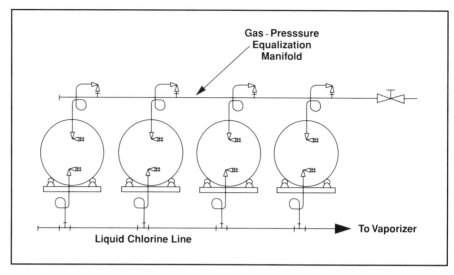

Ammonia

Like chlorine, ammonia is supplied as a liquefied, compressed gas in cylinders, horizontal containers, portable tanks, on-site storage tanks, over-the-road tank trucks, and railcars. Containers must comply with DOT regulations. The most commonly used containers for water treatment plants are cylinders, portable tanks, and on-site storage tanks. The amount of ammonia in each cylinder varies depending on the type of cylinder used.

Cylinders used in ammonia service are different than chlorine cylinders. Filled ammonia containers must provide protection against thermal expansion. Cylinders are filled to 56 percent of volume capacity and are available in weights of 100 and 150 lb (43.3 and 68 kg). Ammonia cylinders have outside diameters ranging from approximately 12 to 15 in. (304 to 381 mm) and thus are wider than chlorine cylinders, which are typically 8 to 10 in. (203 to 254 mm) in diameter. Unlike their chlorine counterparts, neither ammonia cylinders nor ammonia valves are equipped with pressure-relief devices. This is because the cylinder is designed for 480 psig (3,310 kPa [gauge]) minimum pressure. At normal use temperatures of 70°F (21°C), ammonia pressure is 114 psig (786 kPa [gauge]); at 100°F (38°C), the pressure is 197 psig (1,358 kPa [gauge]).

The valves for ammonia cylinders have dip tubes to facilitate the removal of gas or liquid from the cylinders. Dip tubes are small lengths of tubing from the base of the container valve that curve to the container wall. While chlorine liquid feed from cylinders is obtained by inverting the cylinder, ammonia liquid feed is obtained by positioning the cylinder and orienting the valve so that the dip tube is in the phase (gas or liquid) desired. Most water plants feed gas from cylinders and care must be taken so that only gas is discharged. A horizontal cylinder orientation may be necessary to clear the dip-tube inlet of any liquid. It is strongly recommended that the water plant communicate to the supplier that it wants only gas and would prefer to maintain the cylinder in an orientation that permits gas discharge.

Occasionally chlorine-style ton containers are used for ammonia service. However, they do not contain 2,000 lb (907 kg) of ammonia and the valves supplied are suitable for ammonia service only. The amount of ammonia in a ton container is 800 lb (362.8 kg). The ton container used for ammonia contains two valves and is equipped with fusible plugs located at each end. A vertical alignment of the valves is necessary for either gas withdrawal (top valve) or liquid withdrawal (bottom valve).

In some instances, ammonia has been supplied to water treatment plants in portable or "nurse" tanks, which are commonly used in agricultural applications. These tanks have a capacity greater than 1,000 lb (453.5 kg) of ammonia. The user must ensure that gas can be withdrawn from the tanks. Again, contact with the chemical supplier and a discussion of the water plant needs is strongly recommended.

Most ammonia requirements at water treatment plants call for ammonia to be fed at 25 to 33 percent of the amount of chlorine that is fed. As a result of the low consumption, ammonia cylinders are the container most often used by water plants. The continuous withdrawal rate of ammonia gas from 150-lb (68-kg) cylinders is about 40 lb/d (800 g/h) at ambient temperatures of 60 to 70°F (16 to 21°C). Higher rates for shorter time periods may be obtained. If ammonia needs exceed this rate, then a manifold of several ammonia cylinders is recommended. Feed rates that cannot be maintained or condensation and icing are clear indications of inadequate ammonia availability. It may be necessary to use ammonia vaporizers to meet feed requirements.

Physiological Effects of Exposure to Ammonia

Ammonia is not a cumulative poison. The effects of exposure to ammonia vapor depend on exposure concentration and time of exposure. The effects may vary from mild irritation to breathing difficulties. Individuals can have different responses to exposure. Those with chronic respiratory diseases should avoid exposure to ammonia. Table 6-2 provides typical physiological effects.

Reliquefaction

As with chlorine, reliquefaction is a concern with ammonia feed systems. The methods used to minimize reliquefaction in chlorine apply to ammonia as well. These include tracing gas pressure lines, the use of

Ammonia (ppm)	Effect
5	Least perceptible odor
20–50	Readily detectible odor
50–100	No discomfort or impairment for prolonged exposure
150–200	General discomfort
400–700	Severe irritation of eyes, ears, and nose
1,700	Coughing, bronchial spasms
2,000–3,000	Dangerous, may be fatal
5,000–10,000	Serious edema, rapidly fatal
10,000+	Immediately fatal

Table 6-2
Physiological effects of ammonia

Source: *Compressed Gas Association (1992).*

pressure-reducing valves, and drip legs. However, when any electrical devices are to be used, caution must be exercised. Installations involving ammonia must meet the requirements for use in hazardous locations class I, group D, division 1 of the National Electrical Code, Articles 500 and 501 (National Fire Prevention Association 1993).

Materials of Construction

Ammonia is reactive with copper and zinc products and their alloys. The valves used in ammonia service are made of steel. All ammonia pressure piping should be steel, schedule 80 for threaded joints and schedule 40 for welded joints. Flexible connections between storage vessels and manifolds must be of steel construction. NOTE: *Valves and flexible connectors are not interchangeable between ammonia and chlorine service.*

Sodium Hypochlorite Solutions

Since the sodium hypochlorite solutions typically used at water treatment plants are classified as corrosive liquids, there are regulations regarding the shipment of the chemicals. DOT requires that bleach solutions be shipped in corrosion-resistant containers and be provided with pressure-relief devices to prevent overpressurization.

Containers used for shipment or storage may be made from a number of materials, including PE, PVC, and CPVC. Metals should not be used for containment or piping; however, titanium and tantalum have been found to be satisfactory. In some cases, lined steel containers are used. These containers are lined with suitable plastic or other corrosion-resistant materials. Rubber-lined steel or FRP are used satisfactorily for bulk storage. Polyethylene tanks of 5-, 15-, or 55-gal (19-, 57-, or 209-L) capacity are frequently used as shipping containers. Whatever the material of construction, all storage and shipping containers must be equipped with pressure-relief devices. On-site bulk storage facilities are provided for many locations so that delivery by bulk tank truck of up to 4,400-gal (16,720-L) capacity to the storage tank provides for a more cost-effective use. Transfer to on-site storage tanks is usually accomplished by applying sufficient air pressure to the shipment tank to transfer the solution to the storage tank.

All piping materials for sodium hypochlorite should have sufficient chemical resistance and mechanical strength. Lined steel pipe has been used for transfer and feed piping. Lining materials include rubber, PE, and PVC. Although some of the more exotic metals (for example, tantalum and titanium) have been used in piping, the common metals will corrode in the presence of hypochlorite. In fact, metals will accelerate or cause the dangerously rapid hypochlorite decomposition of sodium hypochlorite solutions, emitting oxygen gas. This can cause pressure buildup in piping systems and has been known to rupture plastic piping systems. Such components in the piping system as valves, gauges, and pressure switches must be made of or protected by compatible materials, including PVC, CPVC, and PP.

The wet end of chemical feed pumps must be made of suitable materials. Pump suction intake should be above the storage tank bottom to avoid feeding any sediment into the pump. Storage tanks should be designed so that they can be drained. The tanks should have a vent to prevent pressurization. In addition, a vent is required to prevent the creation of a vacuum during liquid removal and to permit exhaust during filling operations. All tanks must have a suitable level gauge. Translucent storage tanks will provide a visual indication that may be satisfactory. These types of tanks have gauge markings on the side. Translucent tanks should also have ultraviolet (UV) radiation protection to prevent decomposition.

Recent regulations require that all hypochlorite tanks, permanent or temporary, have a containment device, such as a dike or suitable basin large enough to contain the release of all materials in the storage vessel. In many areas, storage tanks that are normally vented must have the vents discharge to a scrubbing device to prevent chlorine off-gas from discharging to the atmosphere (Privetti and Neethling 1993).

Because the solutions of sodium hypochlorite are 80 to 90 percent water, freezing conditions are a concern. Outdoor installations in areas with below-freezing temperatures may require auxiliary heat, not only to protect the tank and piping from freezing, but also to aid pumping. This is because viscosity will be affected by temperature drops. Twelve percent solutions of hypochlorite will freeze at about 5°F (–15°C). Table 6-3 provides some freezing point information. Indoor installation is generally the most satisfactory solution.

Ammonia Solutions

Ammonia solutions are available in strengths up to 25 or 30 percent. They are shipped in bulk containers and then transferred to on-site storage tanks sized to meet the feed rates desired.

Facility Requirements

Each state may have its own storage and use requirements as well as those recommended by the various industry associations (e.g., The Chlorine Institute, CGA, and Chemical Manufacturers Association) and the federal government (OSHA and USEPA) regarding the storage and use of these chemicals. It is incumbent on the user to determine the applicable requirements. In addition, common sense should prevail and any errors should be made in the direction of safety.

Arguably one of the most focused sets of standards for water industry use is that promulgated by the Great Lakes—Upper Mississippi River Board of State Public Health and Environmental Managers, commonly called the *Ten States Standards*. These standards are written specifically for the water industry and are used as the guide for chlorine, ammonia, and chemical feed installations. The *Ten States Standards* was developed by representatives from the following ten states and one Canadian province: New York, Pennsylvania, Ohio, Indiana, Michigan, Illinois, Wisconsin, Iowa, Minnesota, Missouri, and Ontario. The

Table 6-3
Freezing points of sodium hypochlorite solutions

Weight Percent	Freezing point, °F (°C)
2	28.0 (–2)
4	24.0 (–4)
6	18.5 (–7.5)
8	14.0 (–10)
10	7.0 (–14)
12	–3.0 (–19.5)
14	–14.0 (–27)

Source: The Chlorine Institute (1992).

standards are updated from time to time to meet new developments. Some of the requirements from the *Ten State Standards* are as follows:

- A minimum of two feeders shall be provided (one for standby).
- When booster pumps are used, a standby pump shall be provided.
- Separate feeders shall be used for each chemical provided.
- Automatic controls, when provided, shall have a manual override.
- Feeders shall have an electrical interlock with the well or service pump.
- Chemical feed rates shall be proportional to flow.
- Scales shall be provided for measuring the quantities of chemicals used (Figures 6-10 and 6-11).
- Liquid chemical feeders shall have antisiphon protection.
- Water supply to feeders shall have cross-connection protection.
- Chemical feed equipment shall be located in rooms separate from the rest of the facility.
- Space should be provided for 30 days of chemical storage.
- Day tanks should hold no more than a 30-h supply.
- Day tanks shall be provided with a calibrated gauge.
- Respiratory masks of the self-contained, air supply type shall be provided.
- Ammonia solution for chlorine leaks and hypochlorite solution for ammonia leaks shall be provided.
- Chlorine Institute emergency kits shall be provided when using ton containers.
- When a leak detector is provided, it shall have both an audible alarm and warning light.
- Protective equipment consisting of rubber gloves, an apron or similar clothing, goggles or face mask, deluge shower, and eye-washing device shall be provided.
- Chlorine cylinders should be isolated from operating areas, chained in an upright position, and stored in rooms separate from ammonia.
- Chlorine and ammonia rooms shall be enclosed, separated from other rooms, provided with a shatterproof inspection window, provided with doors that have external exits only, and have doors equipped with panic hardware.
- Chlorine and ammonia rooms shall have ventilating fans with a capacity of one complete air change per minute; air inlets shall include louvers near the floor; air discharge shall be above grade, not near walkways, doors, or fresh-air intakes.
- Chlorine and ammonia rooms shall be provided with ventilation.

- Chlorine and ammonia rooms shall be heated to 60°F (16°C) and pressurized chlorine gas shall not be carried beyond the chlorinator room.
- Chlorine rooms shall not be below ground level.

Figure 6-10
Mechanical scales for 150-lb (70-kg) cylinders of chlorine or ammonia are used to provide data on the hourly consumption of actual chlorine or ammonia feed

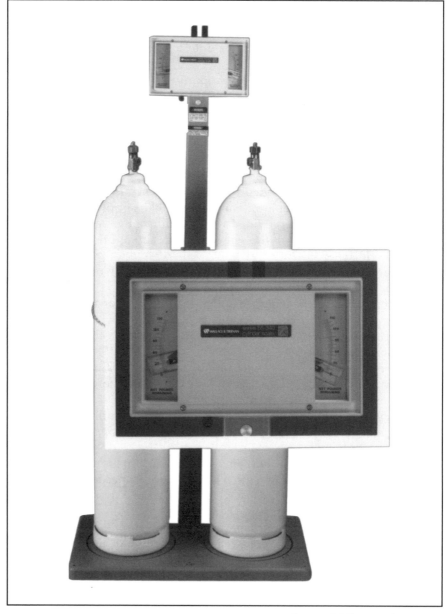

Source: Wallace & Tiernan Inc.

Although this list appears lengthy, it is by no means complete. It is recommended that interested parties obtain a copy of the *Ten State Standards* for review. These standards have no legal standing unless the state codes require their use or have adopted them in some form.

Fire Codes

One of the most recent developments regarding the storage and use of chlorine, ammonia, and their solutions comes from fire codes. These are produced by the various code development organizations as part of a package of codes that includes mechanical, electrical, building, heating, ventilation, etc. These codes have no legal impact until they are adopted or enacted in whole or in part by local municipalities, counties, or states. The local code must be considered when designing and operating water

Figure 6-11
Electronic chlorine cylinder scales with sonic regulators. Electronic scales are available that can be used to transmit a signal to a remote or control area.

Source: Capital Controls Company Inc.

treatment plants. Code use varies from state to state, county to county, and city to city. Each water treatment facility should be aware of the local situation and any applicable codes, and consult with the local fire chief.

There are three major model fire codes currently in use in the United States. They are the Uniform Fire Code, the Standard Building and Fire Code, and the National Code (sometimes referred to as the BOCA Code). The Uniform Fire Code was developed by the Western Fire Chiefs Association (WFCA) and the International Fire Codes Institute (IFCI). The Standard Building Code was developed by the Southeastern Fire Chiefs Association and the Southwestern Fire Chiefs Association. The National Code is a product of the Building Officials and Code Administrators of America (BOCA).

In addition to these codes there are standards developed by the National Fire Protection Association (NFPA) that are used as guides and frequently referenced in the three model codes. In general, the Uniform code is used by states west of the Mississippi River; the Standard code by states south of the Mason–Dixon line, east of the Mississippi, and Texas, Louisiana, and New Mexico; and the National code by states north of the Mason–Dixon line and east of the Mississippi (Figure 6-12). However, these arbitrary, flexible boundaries are just that and political entities choose the code that they desire. It is important to know the code in each locality so that the proper requirements may be understood and met (Privetti 1993).

The three codes have quantity exemptions that permit the use of any of the chemicals (chlorine, ammonia, and solutions of hypochlorite and ammonia) depending on storage and use conditions. Because chlorine is classified as a toxic compressed gas for code interpretations, its requirements are the most rigid of the four. Ammonia is classified as a corrosive compressed gas, and both the chlorine and ammonia solutions (hypochlorite and aqueous ammonia) are classified as corrosive solutions.

Chlorine's exemptions are dependent on the use of fire walls for the storage or use area, as well as the availability of sprinklers. Four 150-lb (68-kg) cylinders of chlorine may be installed in a control area without an exhaust containment system under all three codes, providing the control area has a fire wall with a 1-h rating and sprinklers. Four control areas, each with a 1-h rated fire wall and sprinklers, may be in one building without a containment system. A containment system is generally considered by the codes to mean an exhaust treatment or scrubbing system. The Standard and National codes permit the use of the Chlorine Institute emergency kits in lieu of containment systems for each control area storing or using four cylinders or less. The Uniform code requires a containment system that has been generally interpreted as a scrubber if the exempt quantity is exceeded.

Although not adopted at this writing, one code may accept a shutoff valve at or near the cylinder in lieu of containment systems. All codes

Figure 6-12

Basic national
fire codes

Uniform Fire Code
Standard Fire Code
National Code
Overlap Between Uniform and Standard Fire Codes
Overlap Between Standard and National Fire Codes

require gas detection with an alarm connection to a local fire department or similar public safety facility. The codes are constantly reviewed and updated. They are reissued every three years and have annual updates. It is important to keep abreast of the local code requirements and proposed changes and to review the water plant installation with local fire code personnel on a regularly scheduled basis. Involvement of the local fire department with the emergency planning procedures is strongly recommended. The exempted number of cylinders under each of the three codes and the NFPA standards are as follows: (1) for storage in areas unprotected by sprinklers or a cabinet: 1, (2) within a cabinet in an unsprinklered building: 2, (3) in a sprinklered building, not in a cabinet: 2, and (4) within a cabinet in a sprinklered building: 4. A cabinet refers to a fireproof and airtight storage area.

Under the codes, 810 ft^3 (22.9 L) is considered the volume of a 150-lb (68-kg) chlorine cylinder. One cylinder may be used without any containment. Outdoor storage requires no containment. Table 6-4 indicates the number of cylinders permitted without containment systems. In water treatment plants, the isolated storage and use area of the chlorine or ammonia rooms is considered to meet the definition and intent of a "gas cabinet" since chlorine and ammonia rooms are isolated from the rest of the buildings and are constructed of fireproof materials.

The National and Standard codes accept the use of emergency kits, while the Uniform code requires a containment system that has been generally interpreted as a scrubber. Emergency kits contain materials for capping leaking valves, leaking container walls (cylinder and ton), and fusible plugs. Instructions for each kit's proper use are contained within the kit.

Table 6-4
SARA threshold and reportable quantities

Chemical	Threshold Planning Quantity		Reportable Quantity	
	lb	kg	lb	kg
Ammonia	500	227	100	45.5
Chlorine	100	45.5	10	4.5
Sodium hypochlorite	—	—	100	45.5
Sodium bisulfite	—	—	5,000	2,270
Sulfur dioxide	500	227	—	—
Ammonium sulfate	—	—	—	—

Source: USEPA (1992).

Figure 6-13
Chlorine Institute emergency kit A. Used to stop leaks in cylinders.

Source: Indian Springs Manufacturing Company Inc.

Although some codes may not have additional requirements, The Chlorine Institute recommends, and common sense dictates, that each installation using cylinders have an emergency kit A (Figure 6-13) and a self-contained breathing apparatus on site or readily available. Those with ton containers should have an emergency kit B (Figure 6-14) and a self-contained breathing apparatus on site or readily available. Sites not manned around the clock should have a chlorine detector that is designed to provide an audible or visible alarm or signal a remote, manned location.

Emergency kits are also available for tank cars and tank trucks (kit C) (Figure 6-15) and barges (kit D).

Safety Considerations

Most of the recommendations for a safe installation have been covered in *Ten State Standards* and the fire codes. However, safety also requires training, an understanding of the materials being handled, and a commitment on the part of the utility or water treatment plant. The following items provide the minimum requirements for a safe operation:

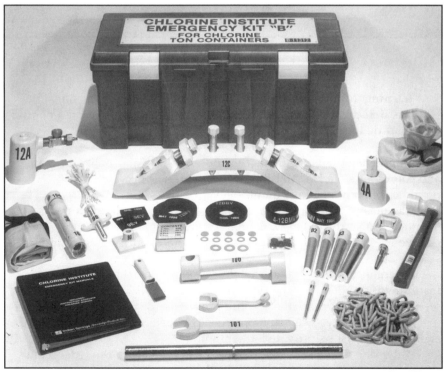

Figure 6-14
Chlorine Institute emergency kit B. Used to stop leaks in ton containers.

Source: Indian Springs Manufacturing Company Inc.

Figure 6-15
Chlorine
Institute
emergency kit
C. Used to stop
leaks in tank
cars and tank
trucks.

Source: Indian Springs Manufacturing Company Inc.

Training. Monthly training sessions with operating personnel.

Gas detectors. Functioning gas detectors that are tested on a regular basis and tie in to an external alarm and an interface with off-facility personnel.

Emergency plans. Plans that include local fire and police personnel and county or state HazMat (hazardous material) personnel with periodic training meetings for all relevant parties.

Gas masks. Appropriate availability and use of self-contained breathing apparatus (SCBA). Escape respirators may be used for emergencies and are sometimes used by operators during entry into storage areas.

Emergency kits. Local availability of the style of the Chlorine Institute kit designed for the type of container in use. Frequent training sessions are recommended to ensure proper use of the kits.

Safety gear. Eye wash and emergency showers (Figure 6-16), gloves, and goggles must be provided at a minimum.

Water department personnel must be familiar with the federal and state requirements for reporting spills under any of the numerous

Figure 6-16
Combination
eye wash and
emergency
shower

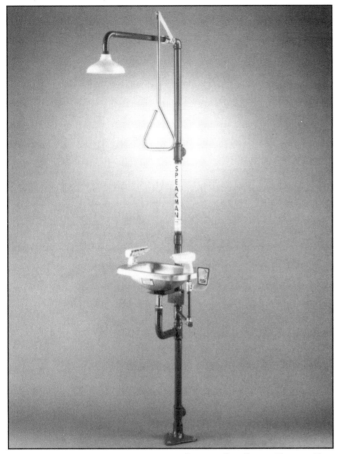

Source: Speakman Corporation.

applicable regulations, including Superfund Amendments and
Reauthorization Act (SARA) Title III. For the chlorine and ammonia
chemicals discussed in this book, Table 6-4 summarizes the threshold
planning quantity, above which emergency planning activities are
required, and the reportable quantities, for which releases in excess of
the amounts shown require reporting subject to state and local reporting
regulations. (Contact your chemical supplier for the appropriate
guidelines.) For emergency needs regardless of the chemical, contact
CHEMTREC at 800-424-9300. CHEMTREC is the North American
emergency information system operated by the Chemical Manufacturing
Association for 24-h assistance during emergencies.

Material safety data sheets (MSDS) are required by OSHA for all
hazardous chemicals used at water treatment plants. MSDS must be

provided by the supplier with or prior to each shipment. New copies must be distributed whenever the information is changed on an MSDS. In addition, regulations require that a master MSDS book on chemicals used or stored be maintained at a convenient location for easy reference and use by all plant personnel. All plant personnel should be aware of its location.

References

American Society for Testing and Materials. 1993. ASTM A106, *Standard Specification for Seamless Carbon Steel Pipe for High-Temperature Service.* Philadelphia, Pa.: ASTM.

Building Officials and Code Administrators International. 1993. *The BOCA National Fire Prevention Code.* Country Club Hills, Ill.: BOCA.

The Chlorine Institute. 1986. *The Chlorine Institute Manual.* 5th ed. Washington, D.C.: The Chlorine Institute.

———. 1992a. *Piping Systems for Dry Chlorine, Pamphlet 6.* 13th ed. Washington, D.C.: The Chlorine Institute.

———. 1992b. *Sodium Hypochlorite, Safety and Handling, Pamphlet 96.* Washington, D.C.: The Chlorine Institute.

———. 1995. *Personal Protective Equipment for Chlorine and Sodium Hydroxide, Pamphlet 65.* Washington, D.C.: The Chlorine Institute.

Compressed Gas Association. 1977. *Anhydrous Ammonia, Pamphlet G-2.* Arlington, Va.: Compressed Gas Association.

Great Lakes—Upper Mississippi River Board of State Public Health and Environmental Managers. 1992. *The Ten States Standards, Recommended Standards for Water Works.* Albany, N.Y.: Great Lakes—Upper Mississippi River Board of State Public Health and Environmental Managers.

LaRoche Industries. 1993. *Ammonia Technical Data Manual.* Atlanta, Ga.: LaRoche Industries.

National Fire Prevention Association. 1993. *National Electric Code.* Quincy, Mass.: National Fire Prevention Association.

Privetti, D.A., and J.B. Neethling. 1993. *New Uniform Fire Code Impacts Disinfection Design, Planning Design and Operation of Effluent Disinfection Systems.* WEF Specialty Conference held in Whippany, N.J., May 1993.

The Southeastern Fire Chiefs Association and The Southwestern Fire Chiefs Association. 1994. *Standard Building Code.* Birmingham, Ala.: The Southeastern Fire Chiefs Association and The Southwestern Fire Chiefs Association.

US Environmental Protection Agency. 1992. *SARA Title III, Consolidated List of Chemicals Subject to Reporting Under the Emergency Planning and Community Right-to-Know Act.* 560/4-92-011. Washington, D.C.: USEPA.

Western Fire Chiefs Association and The International Fire Codes Institute. 1995. *Uniform Fire Code.* Tecumelca, Calif.: Western Fire Chiefs Association and The International Fire Codes Institute.

7

Equipment

This chapter discusses the equipment involved in the two systems of feeding chlorine gas (vacuum and pressure), offers a historical perspective of the process, and presents information relating to materials of construction, reliquefaction, and operation and maintenance (O&M). Equipment for feeding ammonia gas and chemical solutions are examined, along with information about materials of construction and operation. Ancillary equipment with control, residual measurement, or safety implications is also introduced, including chlorine residual analyzers, gas detectors, automatic changeover devices, chlorine and ammonia vaporizers, and expansion chambers.

Feeding Chlorine Gas

Equipment to feed chlorine gas is designed to work either under pressure or under vacuum. By far the most common of the two types operates under vacuum, called the *vacuum-operated solution feed chlorinator*. The name stems from the fact that the equipment feeds chlorine gas only when it receives a vacuum signal and the gas that is fed is mixed with water to form a highly concentrated solution fed to the point of application. The other type of chlorine gas feeder, *pressure-operated gas feed*, operates under the pressure supplied by the gas and feeds gas to the point of application.

The vacuum-operated units offer greater safety in the operation of the equipment and handling of chlorine gas. Such units also provide for greater versatility in the application and control of the dosage.

Vacuum-Operated Feeder Operation

The principal components of a chlorine gas feeder are vacuum regulators, flow indicators, flow controllers, and a venturi. The feeder operates by regulating the flow of chlorine gas by controlling and regulating the vacuum conditions upstream and downstream of an orifice or flow-control device. Most modern chlorine gas feeders use two methods of controlling and regulating the gas flow—constant differential-pressure or sonic flow. Constant differential requires maintaining a constant vacuum differential across the orifice (rate-control valve) by using a differential pressure (vacuum) regulator or downstream vacuum regulator. This regulator maintains a constant pressure (vacuum) drop across the orifice for any given setting of the orifice. When this occurs, the operator can adjust the orifice (the rate-control valve) and be assured that the gas flow will be maintained at the desired setting. The use of a differential-pressure or downstream vacuum regulator corrects for any variation in downstream vacuum that would cause an undesirable variation in gas flow.

Sonic flow requires maintaining a minimum upstream vacuum so that a variation of downstream vacuum will not cause any variation in flow. Sonic flow is so named because the minimum upstream vacuum level required to reach the desired condition is such that the velocity of the gas through the orifice is at the speed of sound—sonic velocity. The formula shown in Eq 7-1 is used to calculate the vacuum level required for chlorine gas at sonic flow (Doolittle and Zerban 1955).

The ratio of the downstream pressure to the upstream, for which the sonic velocity is attained, is called the *critical pressure ratio* (r_c). The critical pressure ratio is a function of the specific heat of the gas (Eq 7-1).

$$r_c = (2/K+1) \exp (K/K-1) \tag{7-1}$$

Where:

$$r_c = \frac{\text{upstream vacuum}}{\text{downstream vacuum}}$$

K = gas–specific heat

For chlorine, K equals 1.335 at 68°F (20°C), and Eq 7-2 is the result.

$$r_c = (2/1.335 + 1) \exp (1.335/0.335) \tag{7-2}$$
$$= 0.539$$

If the upstream vacuum is held at a constant 20 in. water (13.973 psia [96.3 kPa (absolute)]), the downstream vacuum must be 14.6 in. mercury vacuum or 7.531 psia (51.9 kPa [absolute]).

The flow of chlorine gas across an orifice increases as the ratio (P_u/P_d) decreases. Sonic flow of chlorine is reached when this ratio reaches 0.539. When $P_u = P_d$, the ratio is 1 and flow ceases.

Figure 7-1 plots gas flow versus vacuum drop across an orifice. As the vacuum level increases, the gas flow increases, although not linearly. At some point, any additional increase in vacuum level causes no further increase in gas flow. The point at which this occurs is the sonic flow of the gas. For chlorine, sonic flow occurs at approximately 14 in. of mercury vacuum level downstream of the orifice. Sonic flow is important because it represents a simpler form of flow control, has fewer components than other typical functioning systems, and reduces the costs of maintenance and service.

Origin and Development of the Gas Feeder

When the use of chlorine gas in water treatment began almost a century ago, only the crudest feeding methods were employed. This meant that the source of chlorine, usually a cylinder containing the gas, was connected to the point of use by a simple pipe or flexible hose. The shutoff valve on the cylinder was used to regulate and control the flow of chlorine gas. More often than not, the connecting piping was of metallic construction and soon failed due to the highly corrosive nature of the acids formed in water solution. The corrosion problem ceased when such nonmetallic materials as hard rubber were used, but there

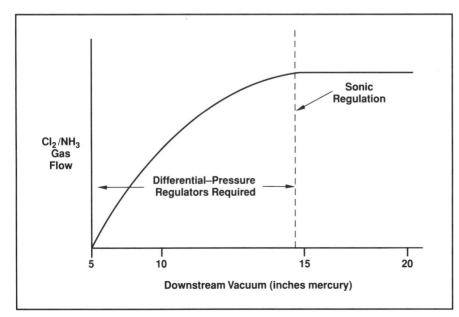

Figure 7-1
Sonic gas flow

101

was no way to determine the quantity of gas being fed and, more importantly from a safety perspective, there was no way to prevent water backup into the cylinder.

As discussed in chapter 6, the pressure in the cylinder varies with the temperature of the liquid chlorine stored in the cylinder. As liquid chlorine evaporates to produce chlorine gas, the liquid temperature decreases and the cylinder pressure drops. With a drop in temperature, the gas feed rate decreases. If the gas pressure decreases below the water pressure at the point of application, then water can flow back through the gas feed pipe or hose and, possibly, into the cylinder.

This undesirable condition would cause corrosion in the cylinder at the chlorine–water interface, resulting in the eventual failure of the container. The release of chlorine from a corroded cylinder would then be a distinct probability. Therefore, to protect the container and control the process, several additional components were necessary. Added to the system were a check valve to prevent the flow of water back to the container and the use of an indicating flowmeter to show the quantity of gas flow. The cylinder valve is meant to be opened or closed and is not designed to be used as a gas-flow control valve. For better control, a gas flowmeter to provide a visible indication of the quantity of gas flowing and a variable orifice to provide for precise, accurate control of the flow of gas were added.

The flow of chlorine gas through the system described to this point is under pressure and feed to the water is limited only by the pressure in the container, the pressure at the application point, and the maintenance of a sufficient differential between them. To compensate for any variations in upstream and downstream gas pressure that could cause undesirable changes in the gas feed rate, monitoring and controlling these pressures is necessary. This is accomplished by the use of pressure-regulating valves upstream and downstream from the control valve. The use of pressure-regulating valves creates a constant differential pressure across the control valve and provides a feed rate that is constant for each setting of the control valve.

The components just described comprise the essence of a direct gas-pressure feed chlorinator. These components are an upstream gas-pressure regulator, an indicating gas flowmeter, a gas rate control valve, a downstream gas-pressure regulator, and a check valve (Figure 7-2).

Although effective in feeding chlorine gas, there are limitations to the use of this type of feeder. Simply put, the gas feeder was restricted to points of addition that were lower in pressure than the pressure in the cylinder. More importantly, the feeder operated under pressure, which is a concern when safety is a consideration. As a result, the pressure-operated gas feeder needed several enhancements. These included the addition of a venturi through which a small volume of pressurized water

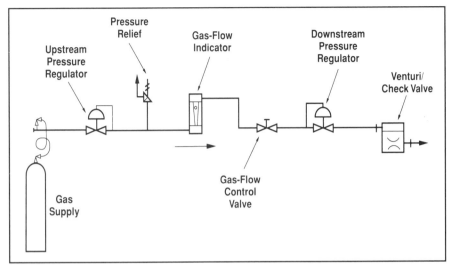

Figure 7-2

Pressure-operated gas feed system

flowed at high velocity. This action created a vacuum at the venturi that acted to open a check valve in the gas piping. To handle the change in the pressure of the gas, the pressure-regulating valves were changed to vacuum-regulating valves. Thus, all the piping and equipment in the gas feeder were under vacuum. The safety of the system was improved considerably. If any of the components or piping broke, the system would fail safe, or, due to loss of vacuum, close the shutoff valve in the vacuum regulator. No chlorine gas would leak from the container into the surrounding area.

Another benefit was obtained from the addition of an ejector. In the ejector, the venturi that created the vacuum now acted as a mixer to form a highly concentrated chlorine solution. Chlorine, which has a low solubility in water, was mixed to form a highly concentrated solution and then added as a solution to the water to be treated. This improved the chlorine's efficiency. Since the chlorine solution blended more easily with the water to be treated and off-gassing was minimized, high-concentration solutions of 700 to 3,500 ppm (mg/L) are typical. These solutions are very acidic, with a pH below 2. Direct gas-pressure feeders do not feed a chlorine solution. The chlorine forms bubbles of gas when added to water. These bubbles accumulate and collect as larger bubbles and increase the probability of off-gassing and escaping from the contact chamber, an undesirable condition.

Finally, the use of the venturi that forces high-pressure water through an orifice could easily be adapted to chlorine addition at higher pressures than contact tanks. Now chlorine could be added to pressure mains by adding higher-pressure booster pumps and by varying the venturi size to meet the desired conditions (Figure 7-3).

There remained only one additional feature to include and that was a relief valve added downstream of the first vacuum regulator. Its function was to relieve or vent any chlorine under pressure that might leak through a faulty inlet or safety shutoff valve in the vacuum regulator.

Modern vacuum-operated, solution-feed chlorinators operate in either mode—using sonic flow control or differential-pressure control. An additional measure of safety is provided by the use of vacuum regulators that mount directly on the cylinder valve or ton-container valve or are used to reduce the gas to a vacuum in the storage area. The vacuum regulators usually include a gas-metering tube, and a pressure-relief and gas-flow control valve. The gas can then be transferred under vacuum from the storage room to the feed control room, which increases the safety of the facility (Figure 7-4).

Induction Mixers

The use of induction mixers (Figure 7-5) is one of the latest developments in feeding gases, such as chlorine or ammonia, and solutions, such as sodium hypochlorite, aqueous ammonia, or other ammonia salts used in water treatment.

These devices enhance the safety of the system design by extending the vacuum line into the mixing chamber (refer to Figure 4-4, page 58) and by eliminating the need for external mixers. By eliminating solution lines, the installation has been simplified and a more rapid response to a feed change is possible. Chemical savings of 20 to 40 percent have been identified by one manufacturer. These savings depend on the process and installation needs and thus are site specific. The use of induction mixers has been extended to other chemicals used in water treatment, for example, carbon dioxide and ferric chloride.

Figure 7-3
Vacuum-operated solution feed system

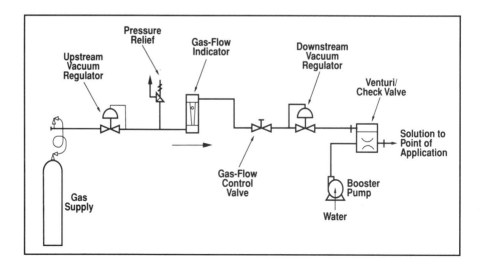

Materials of Construction

The materials used in the construction of gas chlorinators are chosen specifically to handle chlorine gas. The materials are not designed to handle liquid chlorine. All attempts must be made to keep liquid chlorine from entering a gas chlorinator. Major structural components of chlorinators are normally made of acrylonitrile-butadiene-styrene (ABS) or polyvinyl chloride (PVC). Some components may be made of chlorinated polyvinyl chloride (CPVC), which provides some additional features, such as higher temperature resistance. Regulator diaphragms are constructed of Kel-F®, polytetrafluoroethylene (PTFE), or similar fluorocarbons; sealing O-rings of Viton® or similar material; shutoff valves of PTFE and silver; and regulator springs of Tantalloy or Hastalloy C.

All materials must be resistant to corrosion from wet chlorine because there is a possibility that moisture can enter the chlorinator

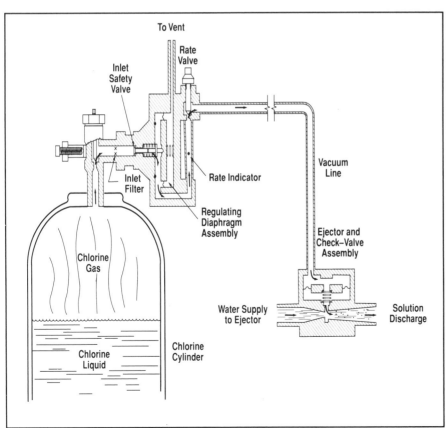

Figure 7-4
Container-mounted sonic vacuum regulator

Source: Capital Controls Company Inc.

through the ejector check valve and form corrosive acids. Thus, the use of silver, Tantalloy, and similar corrosion-resistant metals is preferable. The noble metals (silver and gold) are corrosion resistant to both wet and dry chlorine.

Piping for chlorine gas under vacuum up to 6 psig (41.3 kPa [gauge]) can be of plastic construction, such as PVC, polyethylene (PE), or similar

Figure 7-5
Chemical induction mixers use a high-speed impeller to create a vacuum to draw the chemical, gas or liquid, into intimate contact with water to be treated. Some mixers have an open impeller design (left) and some are closed (right).

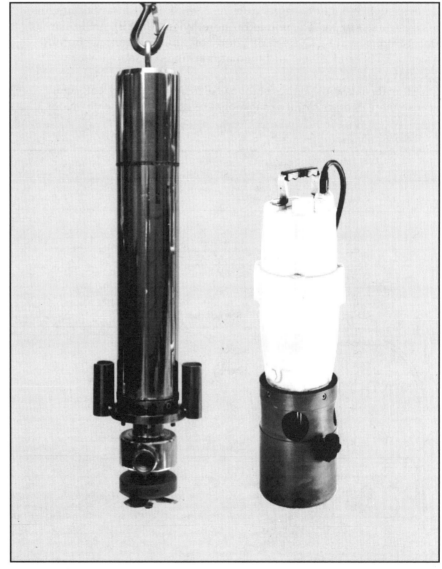

Source: Gardiner Equipment Company Inc.   *Source: Capital Controls Company Inc.*

materials (The Chlorine Institute 1995). Steel pipe, satisfactory for use in dry pressure lines, is not recommended for chlorinator vacuum lines where the presence of moisture can occur if the ejector check valve leaks. If steel pipe is desired for chlorinator vacuum lines, it must be lined with a corrosion-resistant coating. These linings are generally made of PVC or other elastomeric material.

Reliquefaction

As discussed in chapter 6, whenever a chlorine gas feeder installation makes use of gas pressure piping to connect the feeding equipment with the source of chlorine, reliquefaction (the formation of chlorine droplets) in the pressure piping can occur. This is to be avoided at all costs since the gas feeder is designed to handle gas and not liquid chlorine or chlorine droplets.

Reliquefaction will occur when the gas is saturated or at a temperature and pressure such that its physical condition is located at a point on the vapor pressure curve shown in chapter 2 (Figure 2-1, page 11). Any reduction in temperature will cause condensation to occur on the pipe surface, this condensation can be carried as liquid droplets along with the gas stream to the gas feeder. To avoid reliquefaction in gas pressure lines (Figure 7-6), the following techniques are suggested:

- Heat trace gas pressure lines to maintain the pipe surface temperature above the reliquefaction point. Insulation may also

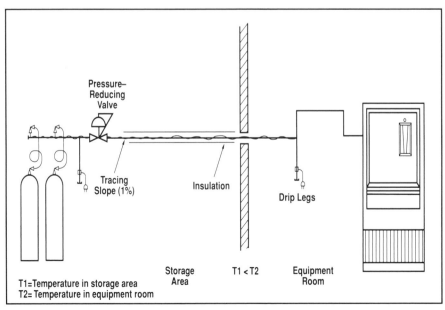

Figure 7-6
Reliquefaction prevention in pressure piping. Temperature in equipment room must be higher than temperature in storage area.

Pressure–Reducing Valve

Tracing
Slope (1%)

Insulation

Drip Legs

Storage Area

$T_1 < T_2$

Equipment Room

T_1=Temperature in storage area
T_2= Temperature in equipment room

be beneficial, particularly if the gas piping passes near or through a cooler area (e.g., outdoors, near a window, underground, or in a basement).

- Piping should never pass from a warmer area to a cooler area. The ambient temperature should always increase in each successive room or area through which the pressure piping passes.
- Piping should always slope toward the source so that any reliquefied chlorine can drain back toward the supply rather than toward the gas feeder. The addition or inclusion of drip legs (down pipes) at points of direction change will help collect any droplets, particularly if the drip leg is equipped with a small (25- to 50-watt) heater.
- Some installations may find the use of a pressure-reducing valve helpful because a drop in pressure will lower the temperature at which reliquefaction occurs.
- The most important recommendation is to install the vacuum regulator at the gas-pressure source or as soon as possible after the line exits the container to reduce the gas to a vacuum (Figure 7-7). When the gas is under vacuum, reliquefaction will not occur until the temperature is below −30°F (−34.4°C). The use of

Figure 7-7

Vacuum piping

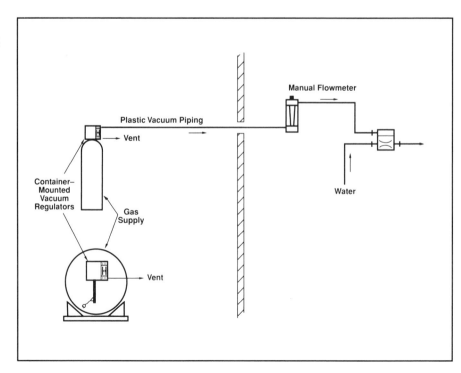

a vacuum regulator mounted directly on a cylinder or ton container valve further reduces the reliquefaction potential because gas pressure lines in the system are eliminated.

- The use of flexible connectors to connect the container to a pressure manifold is frequently practiced. When the gas exits the container under pressure and then enters a pressure manifold, a pressure-reducing valve is recommended as soon as possible in the pressure line in order to minimize reliquefaction. However, reliquefaction can still occur at ambient temperatures. For example, if the container and gas pressure line were 80 psig (551.6 kPa [gauge]), the reliquefaction temperature is 67°F (19.4°C). If the pressure were reduced to 65 psig (448.1 pKa [gauge]), liquefaction would occur at 56°F (13.3°C). Refer to Figure 2-1, page 11.

Service and Maintenance

Each chlorinator manufacturer provides service and maintenance instructions that should be followed for optimum operation. These instructions may vary from manufacturer to manufacturer. However, several fundamental points are common to all manufacturers' manuals. Among these are the following:

- Keep all moisture away from any chlorine system. Any moisture will form acids with chlorine that can cause corrosion, particularly in steel pressure piping. The corrosion products can build up in the piping system and plug the piping so that the flow of gas is severely restricted or stopped. The corrosion products can be carried along with the gas stream to the gas chlorinator. The presence of these materials in the chlorinator will cause blockage in valves and small orifices that will stop gas feed and cause downtime.

- Impurities in chlorine can create feed problems. These impurities may be organic or inorganic in nature. A visual indication of a developing problem can be seen at the gas-metering tube. The tube will darken with brownish red impurities, such as ferric (iron) chlorides from chlorine reactions with the metallic piping, or an off-white, wax-like material, which consists of chloro-organic compounds left in chlorine from the chlorine manu-facturing process. In either case, the chlorinator will have to be taken off line and cleaned. Hot water is normally sufficient to dissolve the iron salts. An organic solvent such as methanol will normally remove the organics. Refer to the manufacturer's instruction manual for further details relative to the particular equipment in use. All impurities tend to collect at points of pressure drop, such as valves and flowmeters. These points must be examined and thoroughly cleaned. Be sure to dry the cleaned components thoroughly before reuse.

- At points of pressure connection, such as at the exit from cylinders or ton containers, it is important to use gaskets that are not brittle. Although frequently found in chlorination installations, fiber gaskets are *not* recommended since they become friable and frequently fray, break off, or disintegrate— particularly after being used several times. The fibers from this type of gasket often find their way into the gas stream and cause pluggage or create flow restrictions in the gas chlorinator. Lead gaskets are recommended at frequently changed connections to provide a good seal with a material that has more desirable physical characteristics.

- The life of a flexible connector, sometimes called a pigtail, can be extended by using isolating valves at locations where frequent connection changes are made. Isolating valves protect the open end of the flexible connector from moisture intrusion during the change. The frequency of container changes and moisture content of the chlorine gas affects the life of the flexible connector. Flexible connectors are normally made of copper tubing with external corrosion protection, usually cadmium plating. Copper tubing is quite satisfactory for this purpose. Internal corrosion is possible in the presence of moisture, but often is not externally visible. However, internal corrosion can be detected by listening for squeaking sounds from tubing movement during container connection changes. When squeaking occurs, replace the connectors. Connectors should be changed once a year or more frequently depending on the number of times containers are changed. Since the flexible connector is under pressure, any indication of a leak must be given prompt attention. Flexible connectors that have been crimped must be replaced.

- Each manufacturer recommends the necessary spare parts to keep equipment operational and recommends an adequate supply of these spare parts, which should be maintained in an appropriate location. Operating and maintenance personnel should ensure that the supply is readily available and avoid the indiscriminate use of a spare or standby feeder as a source for the parts.

Feeding Ammonia Gas

The principles involved in feeding ammonia gas and the equipment used for this purpose are the same as those for chlorine, with the exception of some different operating conditions and materials of construction that result from the chemical and physical characteristics of ammonia. Ammonia feeders are usually called ammoniators. Like chlorinators, ammoniators may either be vacuum-operated solution feed or pressure-operated gas feed. *Do not attempt to use chlorinators as ammoniators and do not attempt to use ammoniators as chlorinators.*

Materials of Construction

Most ammonia gas feeders are constructed of ABS with Kel-F® diaphragms, but the metallic components (springs and control valves) are usually made of stainless steel. O-rings are of neoprene rubber composition.

Operational Considerations

The vacuum-operated unit creates a vacuum in the venturi, just as in the gas chlorinator, but the solution of ammonia and water formed in the venturi is highly alkaline with pH values above 12. Because waters usually contain some level of dissolved calcium and magnesium carbonates, the rapid rise in alkalinity at the venturi changes the solubility characteristics of these dissolved salts and causes them to precipitate in the nozzle and throat of the venturi. The precipitate causes a change in the hydraulic characteristics of the venturi such that the vacuum level created is altered sufficiently to either reduce or stop the feed of ammonia gas. Thus, particularly with well waters, steps must be taken to keep the feeder operational. This can be done in several ways, with varying degrees of success and with varying cost penalties. The methods normally followed for vacuum feed systems are as follows:

- A water softener is installed in the ejector water supply line. The softener reduces the calcium and magnesium contents sufficiently to provide longer venturi life before cleaning is required.
- The addition of a chelating (chemical binding) agent fed to the ejector water supply line just prior to or at the point of ammonia addition will generally produce good results. Frequently, the use of polyphosphates as the chelating agent is effective.
- Timed acid feeds have also been successfully used. The pH change that occurs due to the acid dissolves the precipitated salts and maintains the venturi in satisfactory operating condition.

The three suggested vacuum feed methods are schematically illustrated in Figure 7-8.

The formation of calcium salts creates similar operating difficulties with pressure feeders. Most pressure feeders add gas through a diffuser inserted in the water channel or clearwell. A pH change occurs at the

Figure 7-8
Ammonia
vacuum-feed
systems with
antipluggage
options

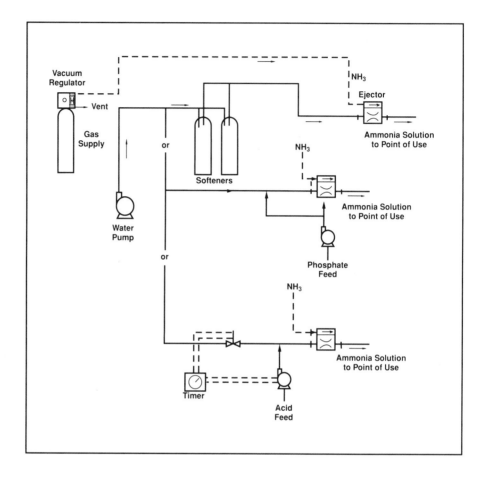

point of addition and a white salt buildup is observed on the diffuser at
the point of gas bubble formation in the water. This buildup clogs the
diffuser and, in time, prevents gas feed. Steps must be taken either to
prevent the buildup or to clean the system. Some methods used are as
follows:

- The use of a neoprene bladder fitted tightly around the diffuser.
  The bladder has extremely fine slits that act as valves and are
  tightly closed when not in use. During use, calcium and
  magnesium salts build up on the slits, which are the ammonia
  addition points, and restrict gas flow. As gas is added, the
  bladder expands, flexing the slits and breaking up the salts. As
  the buildup breaks up, the gas flows into the water. The bladder
  contracts and the cycle repeats itself (Figure 7-9).

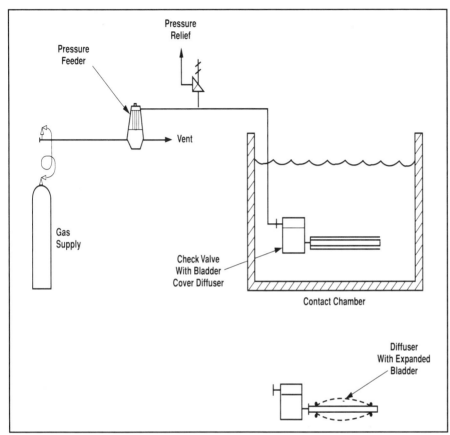

Figure 7-9
Ammonia
pressure gas
feed with
bladder diffuser

- Occasionally, acid washes may be used. This practice is common
  in aeration systems that also experience the salt buildup at the
  aerator nozzles (Figure 7-10).

Gas pressure piping for ammonia pressure lines is usually black
iron, but can be stainless steel. Flexible connectors for use in ammonia
service are constructed of different materials than the traditional copper
tubing used in chlorine service. Ammonia flexible connectors are made
of steel. Valves are also made of steel.

Since ammonia has a higher vapor pressure than chlorine, the
liquefaction problem in gas lines that so frequently plagues the handling
of chlorine may not appear as frequently with ammonia. The liquefaction
preventive measures for chlorine pressure piping would generally apply
to ammonia pressure piping. However, the presence of liquid ammonia
in gas lines can be due to other causes. Typically it is the result of liquid
carry through from the container. It is important, then, that operating

Figure 7-10

Ammonia
pressure feed
with acid wash

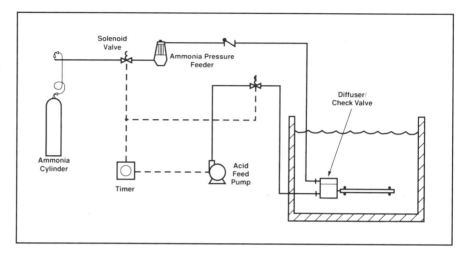

personnel be aware of the type of container provided by the chemical supplier.

Ammonia containers are provided with a dip tube connected to the container's gas-discharge valve. Operators using ammonia containers should determine the type of container and the discharge arrangement by asking the supplier for container information. Discussions with the ammonia supplier are strongly recommended to minimize the presence of liquid ammonia.

Chemical Solution Feeders

Solutions of sodium hypochlorite, aqueous ammonia, and ammonium salts are easily fed using chemical feed pumps. If the feed rate volume is too large for chemical feed pumps, centrifugal pumps may be more economical.

Chemical feed pumps with mechanically or hydraulically controlled diaphragms using solenoid, piston rod, or plunger drives that adjust stroke lengths or vary cycles or motor speed are the types most commonly used. Whatever the type, chemical feed pumps are positive displacement, volumetric pumps that increase the pump chamber volume to receive the chemical to be pumped and then decrease the pump chamber volume to push the chemical to the point of use. A series of inlet and outlet check valves control the fluid movement. In terms of operation, chemical feed pumps are typically more demanding than centrifugal pumps.

Chemical feed pumps with mechanical diaphragms (Figure 7-11) have the drive rod directly connected to the diaphragm. Although these pumps can be very economical, they are also prone to diaphragm failure

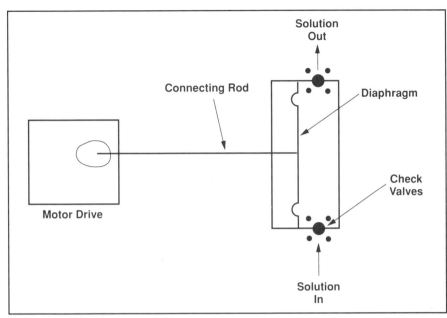

Solution
Out

Connecting Rod

Diaphragm

Check
Valves

Motor Drive

Solution
In

Figure 7-11
Chemical feed
pump mechan-
ical drive

due to the excessive strain placed at the diaphragm/drive-rod
connection point. Solenoid-operated pumps are usually mechanically
connected to the diaphragm and are often used for low flows and low
injection pressures.

Hydraulic drive pumps overcome this action on the diaphragms by
using a hydraulic fluid on the drive-rod side of the diaphragm. The fluid
transmits the force of the drive rod or piston uniformly across the entire
area of the diaphragm and eases the strain on the diaphragm (Figure
7-12).

Some recent pump designs use magnetic drives. Each pump type
must be evaluated based on its features and the application needs before
a choice is made.

Operational Considerations

Pump calibration is an important factor when using chemical feed
pumps. The frequency of calibration and the accuracy required may
dictate that a system of calibration be permanently installed in the piping
system. The simplest way to calibrate a pump is to feed to a graduated
cylinder using a stopwatch (Figure 7-13). Since the chemical feed pump
is a volumetric feeder, the specific gravity of the chemical being used
must be taken into consideration.

Also of importance is the pressure under which the system operates.
Since most chemical feed pumps use small-diameter lines (1/2 in.

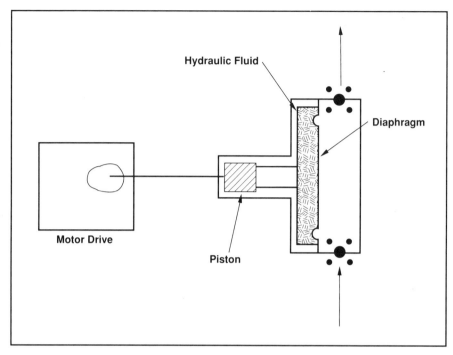

Figure 7-12
Chemical feed
pump hydraulic
drive

Hydraulic Fluid

Diaphragm

Motor Drive

Piston

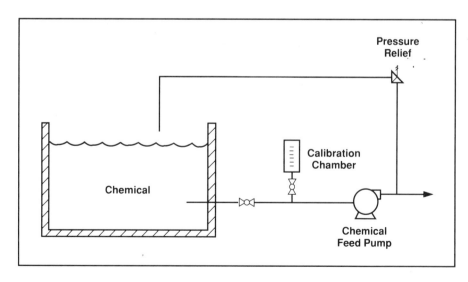

Figure 7-13
Chemical feed
pump
calibration

Pressure
Relief

Calibration
Chamber

Chemical

Chemical
Feed Pump

[13 mm] or less), long pumping distances allow for the possibility of water hammer. High pressures at the point of application are also a concern. Piping must be appropriately designed. A solution to either of these conditions is to install an expansion chamber or pulsation dampeners in the line and review the hydraulic design to consider larger line sizes.

Maintenance and Service

The area of most concern with chemical feed systems in general and hypochlorite feed systems in particular is the buildup of deposits at the inlet and outlet valves. Deposits around the valve seat prevent either valve from seating and lead to a drop-off in feed or complete failure of feed. Valves should be inspected frequently for such buildups. Acid cleaning or water washing is usually sufficient to remove deposited materials. On each such occasion, the O-rings and gaskets used at the check valves should be replaced.

Replacement diaphragms should be readily available. The frequency of replacement will depend on operating conditions. Closing down-stream valves while the pump is operating should be avoided as this will cause a high backpressure and the diaphragm will fail and perhaps rupture unless a relief system is provided (Figure 7-12). The type of diaphragm used and its chemical resistance to the material being pumped is also important. The manufacturer should be consulted on the proper selection of diaphragms. Critical feed equipment should have standby pumps available. Replacement parts must be readily accessible, and people must be trained in the maintenance of the pumps.

On-Site Hypochlorite Generation

There is an increasing interest in generating sodium hypochlorite on-site due to safety considerations, simplicity of operation, and the ready availability of the raw materials required.

Sodium hypochlorite generated on-site is produced in an electrolytic cell using a brine (sodium chloride) solution feed stock (Figure 7-14). As the solution is passed between two electrodes, a direct current is applied. The energy imparted produces chlorine gas, sodium hydroxide, and hydrogen gas as a by-product (Eq 7-3). If there is a membrane between the electrodes, the chlorine gas and sodium hydroxide remain on opposite sides of the membrane. Industrially operated chlorine/caustic systems use the same principle of operation, only on a larger scale, to produce tons of chlorine and caustic. If there is no membrane, the two chemicals combine to form sodium hypochlorite solution (Eq 2-1, page 15). The concentration of the solution produced is approximately 1 percent.

$$2NaCl + 2H_2O + \text{Electric current} \rightarrow Cl_2 + 2NaOH + H_2 \qquad (7\text{-}3)$$

A solution with this low a concentration does not decompose as rapidly as the industrially available concentrations of 12 to 13 percent used at many facilities (see Figure 2-5 on page 19). With low concentrations formed and the slow decomposition that results from these low concentrations, the production of sodium chlorate is

Figure 7-14

On-site hypochlorite generation systems use only salt, water, and electric power to generate sodium hypochlorite. The systems are used at wellheads and treatment plants.

Source: ELTECH International Corporation.

considerably reduced. The raw materials used (salt, water, and electric power) are environmentally acceptable and offer no safety hazard compared to chlorine gas and sodium hypochlorite solutions.

Equipment used includes a water softener, storage tanks for the hypochlorite solutions produced, a brine saturator, an electrolytic cell, and an electrical rectifier. Some chemical feed pumps and instrumentation are also necessary.

Operating expenses depend on power, water, and salt costs. These variables must be determined locally since they are a function of local availability and are site specific. Capital costs depend on the type of equipment installed. For example, if locally available softened water is used, a water softener is not required. A flow schematic is shown in Figure 7-15. The solution produced is fed to the point of application with a chemical feed pump.

Ancillary Equipment

There are many pieces of equipment used in conjunction with chlorine or ammonia gas feeders and hypochlorite or ammonia solution feeders. Only equipment with control, residual measurement, or safety considerations will be discussed here.

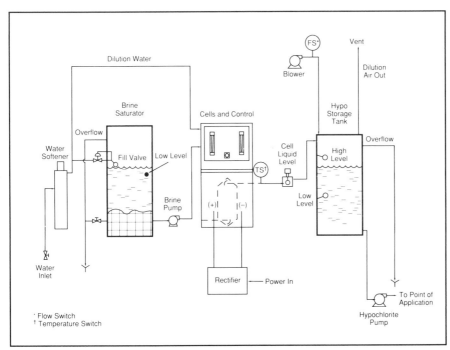

Figure 7-15
On-site hypochlorite flow diagram

Source: Capital Controls Company Inc.

Chlorine Residual Analyzers

Of all the ancillary items used in water treatment systems, the most important is the chlorine residual analyzer. The enactment of the Surface Water Treatment Rule (SWTR) dictates the continuous measurement of chlorine residual in the water. Depending on the size of the utility, periodic sampling may be satisfactory. However, most prudent water utilities are following a course of continuous measurement and are considering control of chlorine feed to meet a preset value. The function of the analyzer is to measure the chlorine residual in the water so that an indication of the residual is available on a continuous basis and a signal can be sent to a controller to pace a chlorinator. Generally the most commonly used analyzers are one of two types: amperometric or colorimetric.

An amperometric analyzer (Figure 7-16) typically consists of two electrodes that are immersed in a continuous water sample. The electrodes are made of two dissimilar metals, e.g., gold and copper, that measure a change in current flow between them that is directly proportional to the amount of chlorine residual in the water stream (Figure 7-17). While most units use the electrochemical potential between the two metals as the driving force, some may have low voltage impressed across the electrodes.

An amperometric analyzer can be used to measure either total or free residual chlorine. For free chlorine, a buffer is added to maintain a constant pH of approximately 4 to enhance signal stability. An iodide salt and buffer are added to permit the measurement of total chlorine residual. The iodide reacts with both the free and combined chlorine residuals to permit measurement of both species. The electrical signal produced at the electrodes is converted to a 4- to 20-milliampere (mA) analog signal for local or remote indication and recording or fed to a controller to be used for residual or compound-loop control.

Colorimetric analyzers consist of a photoelectric cell and a light source that detects a variation in color produced in the sample stream on addition of a reagent specific for chlorine, free or combined. The reagent will react with the residual chlorine to produce a color whose intensity will vary with the amount of chlorine present. The color change is measured in the cell and a conditioned signal is sent to an indicator, recorder, or controller. Most colorimetric analyzers, although indicating continuously, sample on a timed basis and the indicated value of chlorine is changed on a timed-cycle basis. The cycle depends on the reaction time of the chemical and the time to reach a stable color.

A recently developed instrument uses a probe that can be positioned permanently in the water to be sampled or held in a stilling well to which the sample is pumped. These units contain electrodes suspended in an electrolyte and separated from the sample by a membrane. As

Figure 7-16
Amperometric chlorine analyzers use electrodes made of two dissimilar metals to measure free or total chlorine in the water stream

Source: Wallace & Tiernan Inc.

Source: Capital Controls Company Inc.

chlorine molecules pass through the membrane, the amount of current produced varies directly with the chlorine present.

As with any analytical instrument, calibration, service, and maintenance are of primary importance. Chemicals must be maintained in a fresh and ready state. Cleanliness of the measuring cell is critical. Most amperometric devices use in situ cleaning devices, such as small plastic balls permanently located in the cell or periodically added abrasives that are pushed against the electrode surfaces and agitated to maintain a constant state of cleanliness. A colorimetric device generally requires flushing with water or cleansing fluid and periodic cleaning of the photoelectric cell's outside surface.

Periodic calibration of all types of chlorine residual analyzers is recommended. It is a good practice to do this weekly. Many plants routinely check calibration daily. There may be occasions when a situation at a site would dictate a more frequent calibration. Signal drift may be an indication that the measuring cells require cleaning. A calibration check after each cleaning is also recommended. At a minimum,

Figure 7-17

Amperometric analyzer

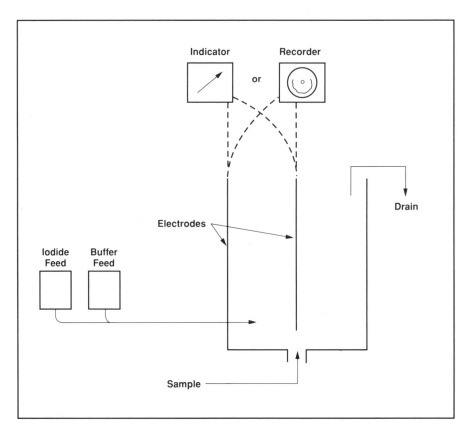

the recommendations of the manufacturer should be followed.

In all cases, the cleanliness of the sample is of utmost importance. Filters or flushing Y-strainers will aid in keeping the cell clean. These same filters and strainers will also require servicing to ensure the sample is not becoming contaminated (Figure 7-18).

Gas Detectors

Devices that measure a gas in the surrounding atmosphere are called gas detectors (Figure 7-19). Federal Occupational Safety and Health Administration (OSHA), state, and local safety regulations require all installations to have gas detectors in operation that are sensitive to levels of gas concentrations in the range of the maximum allowable concentrations (MAC). All storage and use areas should have active gas detection systems in operation. It should be remembered that a detector will not locate the leak but will simply tell the user that gas is present. It is up to the user to respond to the alarm condition and to locate and determine the nature of the leak. The recommended sensitivity levels for chlorine and ammonia gas detectors are both 1 ppm/volume.

Detectors are available for both chlorine and ammonia. The great majority are of the electrochemical type and are capable of detecting small concentrations of gas in the surrounding air by a change in the conductivity of probes that are immersed in an electrolyte. The electrolyte will absorb the chlorine or ammonia and indicate the change as an electrical signal.

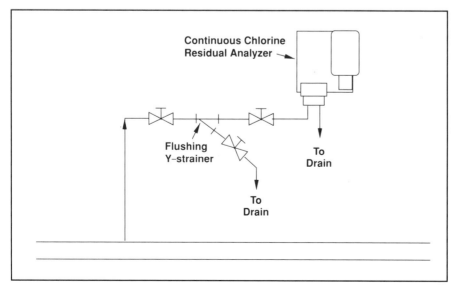

Figure 7-18
Sample line

123

Effective use of the signal is important. Activation of a local external or control room signaling device (e.g., a horn or light) is recommended and a connection to the local emergency planning operation or hazardous material (HazMat) organization may be required by local regulations. The local fire departments usualy respond to these emergencies. Most local fire codes require that an emergency plan be submitted, approved, and coordinated with the local group. Some fire codes may require the use of exhaust scrubbing systems, but organizations such as The Chlorine Institute recommend a risk analysis be conducted to determine the need.

A risk analysis considers such factors as quantity in use; proximity to other facilities, residences, and factories; prevailing winds; availability of trained personnel; and preparation of emergency plans to evaluate the need for containment or treatment systems.

Most current code requirements will permit the storage and use of up to four 150-lb (68-kg) cylinders in a control area if the rooms are sprinklered and constructed with materials suitable for a 1-h fire wall rating. When meeting this requirement, the facility may operate without the need of a containment system (scrubber). In locales covered under

Figure 7-19
Chlorine and ammonia gas detectors act as safety devices by alerting personnel to the presence of the gases in the atmosphere. One indicator can be used to monitor up to eight sites.

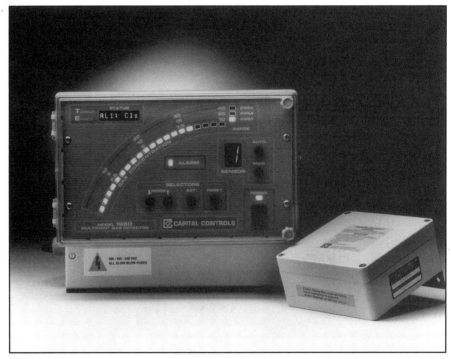

Source: Capital Controls Company Inc.

the Standard or National (BOCA) codes, automatic shutoff devices, emergency kits, containment vessels (coffins), or other containment methods can be employed in lieu of scrubbers. It is important to determine what the local code requirements are relative to these matters.

Automatic Changeover Devices

Most state regulations require 100 percent equipment standby for chlorination facilities, and prudent design and operation should provide the same for ammonia. In addition, the ability to automatically change from an empty supply source to a full standby supply is often required. The use of automatic changeover devices provides the treatment plant operator with some assurance that water will not be sent to the distribution system without a chlorine residual.

Automatic changeover systems (Figure 7-20) are available to either mechanically or electrically change from one source to another. Many installations require the ability to switch gas feed sources. The majority of mechanical devices are vacuum operated and use the high vacuum level or vacuum difference developed when the gas supply is depleted to switch from the empty, on-line source to the full, standby source. These sources may either be cylinders or ton containers. Larger installations that use liquid chlorine or liquid ammonia from ton containers may also

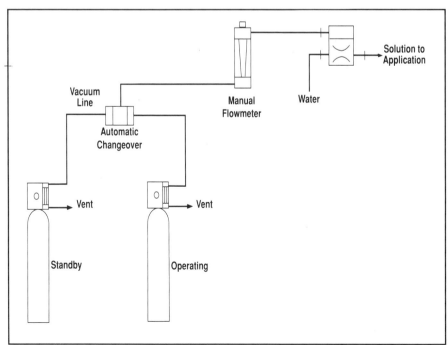

Figure 7-20
Automatic changeover system, vacuum operated

125

desire to switch from one liquid supply source to the another (Figure 7-21). The use of pressure sensors and suitable electrical interlocks provide the necessary capability. Switching from tank car to tank car can be done, but is limited due to US Department of Transportation (DOT) regulations requiring manned attendance not only during operation but also during switching. This usually defeats the purpose of automatic changeover. Manual changing of tank cars is the industry recommendation as the safest procedure.

Chlorine and Ammonia Vaporizers

A vaporizer (Figure 7-22) uses an external heat source to convert liquid chlorine or ammonia to gas at a desired rate. Vaporizers are used at large installations that require gas in quantities greater than that capable of being produced by the supply without an external heat source. In general, when the chlorine feed rate exceeds 2,000 lb/d (40 kg/h), or chlorine is only available in tank cars, trucks, or on-site storage and when the manifolding of multiple sources in a sufficient number to permit gas withdrawal is not possible nor practical, chlorine vaporizers must be used. When the desired feed rate of ammonia exceeds 1,000 lb/d (20 kg/h), or the ability of the source to provide sufficient heat to meet the gas flow rate desired cannot be met, ammonia vaporizers must be used.

Figure 7-21

Liquid chlorine changeover system

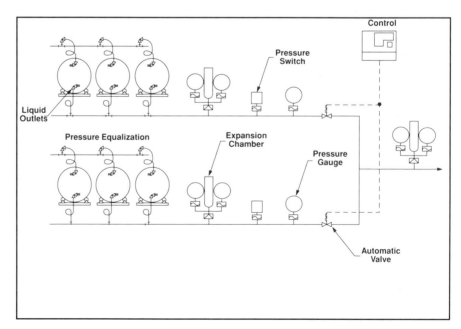

The operation of chlorine or ammonia vaporizers does not increase the pressure of the chemical in either the liquid or gas portion of the system. However, since vaporizers operate under pressure, all vaporizers must be designed to meet the American Society of Mechanical Engineers (ASME) code (1995) for pressure vessels. In addition, since ammonia is

Figure 7-22
Vaporizers are used to convert liquid chlorine and ammonia to the gaseous state for use in gas feeders

Source: Capital Controls Company Inc.

classified as a class 1, group D, division 1 gas, there are National Electrical Code requirements specific for ammonia. Thus, additional evaluation of the installation and provisions in the equipment design are required.

Chlorine and ammonia vaporizers normally have their vaporization chambers immersed in a water bath with the heat being supplied by electric heaters. Some types have an external heater and a pump to circulate the heated water from the heater to the vaporizing chamber. Alternatively, low-pressure steam vaporizers may also be used, although their presence in the water industry is not as common as the electrically heated type. A typical electrically heated, water bath vaporizer is illustrated in Figure 7-23.

Vaporizers consist of a pressure vessel, containing liquid chlorine or ammonia, surrounded by water. The liquid chlorine or ammonia absorbs heat through the pressure vessel walls until the liquid is sufficiently heated to vaporize and exit the chamber. There is neither a chlorine nor ammonia liquid level control because the vaporizer level is maintained by the feed rate established by the downstream gas feeding device. Liquid carryover can occur whenever the feed rate exceeds the ability of the vaporizer to provide sufficient heat or there is a buildup of dirt left in the chamber over periods of vaporization.

The vaporizer must be equipped with several devices for safe operation, including the following:

- Pressure-reducing and shutoff valve (PRV). This valve will close on either low water temperature, low water level, or loss of electric power. The PRV also functions as a device to prevent liquid carryover because the drop in pressure across the valve provides superheat and protects the downstream feeder from liquid chlorine or ammonia.
- Pressure-relief valve. The ASME Pressure Vessel Code requires a pressure-relief device to protect against overpressurization of the vaporization system.
- Temperature control. Both the water bath and the piping for gaseous chlorine or ammonia will have sensors to determine the respective temperatures for control or monitoring purposes. In the case of the water bath, the water temperature is controlled in a narrow range sufficient to vaporize the chlorine or ammonia. High water temperature will shut off the power to the water heaters and shut the system down. The temperature of the gas exiting the chamber must be monitored to ensure sufficient superheat is obtained to minimize the possibility of downstream reliquefication and to provide an indication for scheduling chamber cleanout. Comparing the off-gas temperature from the vaporizer to its position on the chlorine or ammonia vapor pressure curves can be used to identify sludge buildup in the

vaporizing chamber. Ten to 20°F (5 to 10°C) of superheat is considered sufficient. Less than that is an indication of possible sludge buildup or heater failure.

• Water-level switch. A sufficient level of water must be maintained so that the heat generated by the electrical heaters is transferred to the water bath and then to the liquid chlorine for vaporization. The presence of the water surrounding the

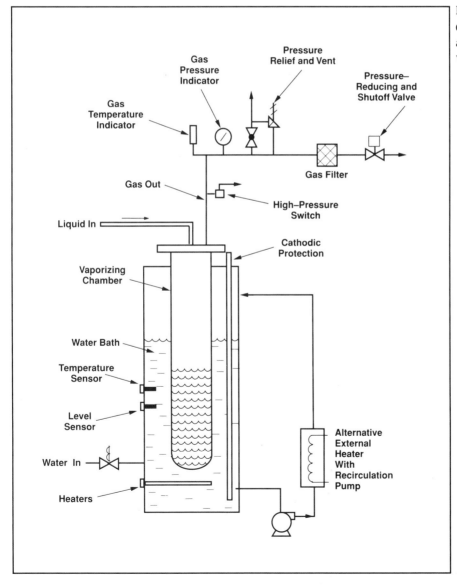

Figure 7-23
Chlorine/
ammonia
vaporizer

chamber acts to ensure that the temperature of the chamber and chlorine are maintained below 180°F (82.2°C). A low water level in the water bath will shut the system down and turn off electrical power.

- Cathodic protection. Protection against corrosion of the vaporizing chamber and the water bath tank is provided on the water side by donor electrodes. An indicating ammeter is used in an electrical circuit to monitor protection and the presence of the electrodes.
- High-pressure switch. In the event of high pressure, the high-pressure switch disconnects the vaporizer heater power supply and provides an alarm contact to permit an external alarm signal.

The Chlorine Institute can provide written information on the satisfactory design, proper installation, safe operation, and recommended minimum maintenance of vaporizers (The Chlorine Institute 1995b).

Expansion Chambers

All liquid chlorine lines must be provided with expansion chambers to protect the liquid line from over-pressurization (Figure 7-24). An expansion chamber must be installed in a liquid line between any two valves. The chamber must be sized to protect up to 20 percent of the line volume. All expansion chambers are equipped with rupture disks that relieve into the empty chamber. A pressure gauge and pressure switch are located between the rupture disk and the expansion chamber to provide a visual indication of the pressure in the chamber and a contact closure that would activate an alarm signal to alert operating personnel of a potentially unsafe situation. The Chlorine Institute's Drawing 183 (The Chlorine Institute 1994) dealing with expansion chambers should be reviewed (also refer to Figure 7-21).

Safety

In addition to the gas detectors for both chlorine and ammonia, there are several installation recommendations relating to the use of some ancillary equipment that should be followed. They include the following:

- Facility design. Chlorine and ammonia gas containers should be kept in separate rooms or, at the very least, be separated by a partition during both storage and use. All doors from the chlorine and ammonia rooms should exit only to the outside. No doors should be provided from the storage and use rooms to any interior rooms or offices. Hypochlorite and ammonia feed facilities should be in separate rooms. All rooms should have exhaust systems designed for one air change per minute.

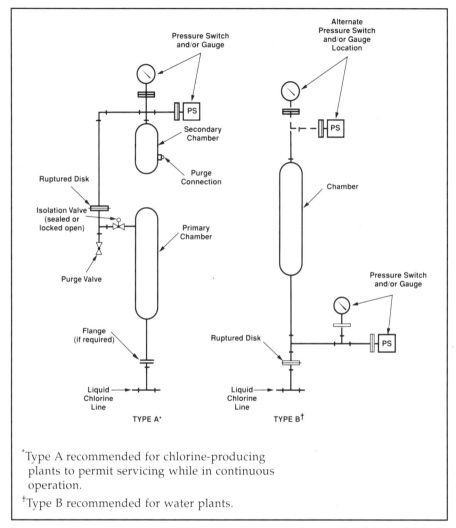

Figure 7-24

Expansion chamber arrangements

Source: The Chlorine Institute.

* Self-contained breathing apparatus (SCBA). At least one SCBA should be housed outside the storage and use room entrance. A spare should be maintained in a convenient but remote location.
* Chlorine Institute emergency kits. These kits should be readily available in a convenient location but removed from locations that would make them inaccessible during times of need. The kits are available for the following three chlorine containers: kit A for 150-lb (68-kg) cylinders, kit B for ton containers, and kit C for tank cars and over-the-road tank trucks. Regular training of

all personnel in the use and maintenance of these kits is recommended for enhanced safety.

- Spill containment. Spill containment may be a requirement of local building codes. Regardless of whether mandated or not, it is a good environmental and safety practice. Tanks, ton containers, and other storage vessels may require diking. Containment design requires the diked area to have no drains, a dike volume sufficient for each diked vessel, and separate containment provided for different chemicals.
- Emergency plan. An emergency plan should be well thought out and posted. All local emergency groups, such as fire departments or HazMat teams, should be involved in the planning.

Operating and Maintenance Costs

There are two cost categories associated with the feeding of chlorine and ammonia gases as well as solutions of hypochlorite and ammonia salts. These are capital costs and O&M costs. Capital, or initial, costs vary with the specifications and needs of the prospective user and are subject to local conditions. O&M costs vary with type of equipment.

Manufacturers provide a list of recommended spare parts for chlorination equipment. It is generally a good policy to keep these parts at a minimum. Personnel must be trained in the maintenance of the equipment. Most manufacturers will conduct training sessions for maintenance and service of their equipment. In addition, some manufacturers provide an exchange program to permit the servicing of their equipment at their facilities. This way the operating personnel can send equipment in for repair and install a spare or exchanged unit for operation during the time of service.

References

American Society of Mechanical Engineers. 1995. *Boiler and Pressure Vessel Codes.* Section VIII, Rules for Construction of Pressure Vessels. New York: ASME.

Doolittle, J.S., and A.H. Zerban. 1955. *Engineering Thermodynamics.* 2nd ed.

The Chlorine Institute. 1995a. *Piping Systems for Dry Chlorine.* 13th ed. Washington, D.C.: The Chlorine Institute.

———. 1995b. *Chlorine Vaporizing Systems.* 4th ed. Washington, D.C.: The Chlorine Institute.

———. 1994. *Manifolding Ton Containers for Liquid Chlorine Withdrawal.* Washington, D.C.: The Chlorine Institute.

Control of Gas and Liquid Feeding Equipment

It is important to review the different methods of control for both gas and solution feeders, the applications that are more appropriate for the particular method, and an explanation of why these applications are appropriate. The equipment involved in both manual and automatic control operates by the same principles, although modifications must be added to permit automatic control. The major modification is in the controller. The controller is an electronic instrument that receives the signal or signals from the sensing instruments. The signals are electronically multiplied or combined to provide one signal to the feeding equipment. Electronic control is the industry standard in modern water treatment plants. As recently as the early 1970s, many plants were still equipped with pneumatic controls. Because of their simplicity, economy, and proven performance, electronic control schemes have replaced most (if not all) of the outmoded pneumatic controllers.

This chapter will not deal with the various controllers available but will review the equipment used for feeding the chemicals, including manual, semiautomatic, automatic proportioning, automatic residual, and compound-loop control gas feeding equipment.

Manual Control Gas Feeder

Manual control means that startup, shutdown, and feed-rate adjustments are all done by hand by the treatment plant operator. Figure 8-1 shows two types of manual feed chlorinators for direct container mounting (cylinder or ton). Startup is accomplished by opening a water valve to the venturi in the injector or starting up a water pump to feed the venturi. The cylinder valve must also be opened to permit gas to flow. The gas feed rate is adjusted to the desired value by turning the rate valve and adjusting the rate by observing the gas flowmeter or indicator. The gas flowmeter will read in pounds per 24-h day or in grams or kilograms per hour. The amount of gas to be fed is determined from the dosage desired and the water flow rate (Figure 8-2). A typical calculation is presented in appendix A. Once the feed rate is established, the equipment operates continuously at that rate until the feeder is either shut down or the feed is adjusted to another rate.

Applications appropriate for manual control would include situations in which water flow and demand are constant and an operator is readily available to make any necessary adjustments. A typical application of this type is in the treatment of water in swimming pools. This type of application should never be left unattended.

Semiautomatic Control Gas Feeder

Semiautomatic control is sometimes referred to as start/stop or on/off control (Figure 8-3). In this control scheme, the operator sets the desired chlorine feed rate manually at a fixed value. Startup and shutdown are done automatically. An electrical signal is supplied from an external source (e.g., a well-pump starter) and is used to activate the gas feeder by opening a solenoid valve and/or starting a booster pump in the water supply line to the venturi in the chlorine injector.

Applications appropriate for semiautomatic control are water systems pumping from underground wells. There is usually no additional treatment of the water provided nor needed. The water has a fairly constant demand. Semiautomatic control is perhaps the most widespread use for chlorination equipment due to the simplicity of the system needs, the safety of the installation, and the ease of adaptability to water systems (Figure 8-4).

Automatic Proportioning Control Gas Feeder

Sometimes referred to as flow proportioning control, this type of control may be the most frequently used for automatic operation. In instrumentation terminology, flow proportioning control may more properly be called *open- or single-loop control*. In open-loop control there is

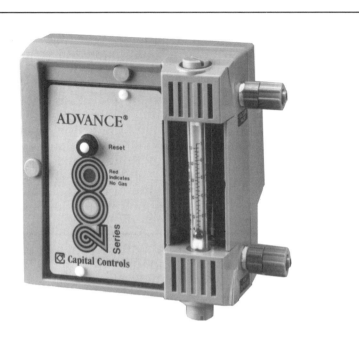

Figure 8-1

Manual feed chlorinators for direct container mounting (cylinder or ton) with capacities up to 21 lb/h (10 kg/h) are used for fully manual or semiautomatic applicators

Source: Capital Controls Company Inc.

Source: Wallace & Tiernan Inc.

Figure 8-2

Manual
chlorination/
ammoniation
system

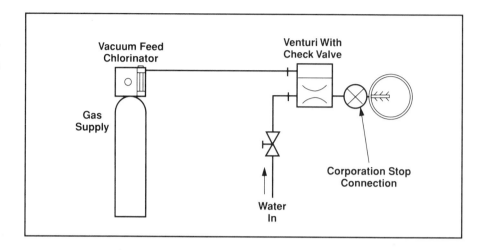

Figure 8-3

Wall-mounted
automatic gas
control valve

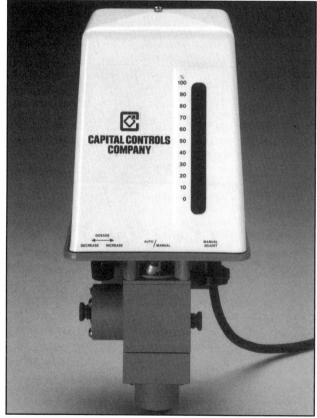

Source: Capital Controls Company Inc.

no signal or feedback to the control system, hence the open loop. This is sometimes referred to as a wild stream control.

In flow proportioning control, two additional items are added to the chlorination system. These are a motorized valve and an electronic controller. The valve used is a linearized valve set so that each incremental change provided as input results in a constantly proportioned change in the valve opening. The chlorination equipment is equipped with an electronic controller that receives information from the sensing instrument (water flowmeter), sends the information into the electronic controller, calculates the desired output based on the desired chlorine dosage, and sends a signal to the valve drive motor to open or close the valve the correct amount. The valve must be equipped with a drive device that will open or close the valve as instructed by the automatic controller. The flow signals are normally electronic, (4–20 milliamperes, direct current [mADC]; 0–20 mADC; or 1–5 volts, direct current [VDC]),

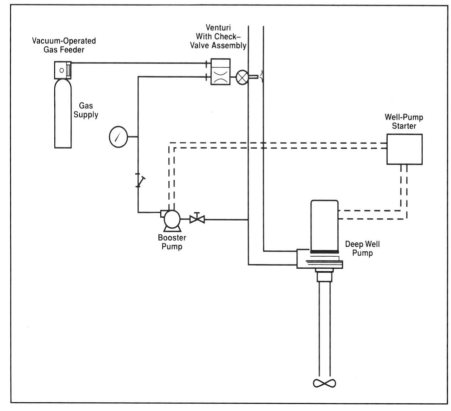

Figure 8-4
On/off feed
system

and the controller will provide the appropriate electrical output to position the valve in direct proportion to the flow signal, hence the name flow proportional control.

Because the chlorine gas flow is altered only in response to variations in water flow, the chlorine demand in the water to be treated must be constant. The dosage desired by the operator must be set manually to meet the desired residual. The results must be measured manually or indicated continuously with a continuous chlorine residual analyzer. There is no feedback to the control system, so the operator must be alert to any deviations from the desired result and respond promptly to make the corrections manually.

Applications for flow proportioning control must be (1) where the demand is constant and only the flow will vary or (2) where deliberate, periodic operator input is desired as a part of routine operation to check residual conditions. The water system must have a means of measuring the flow and providing the 4- to 20-mA electrical signal to the gas

Figure 8-5

Flow proportioning system

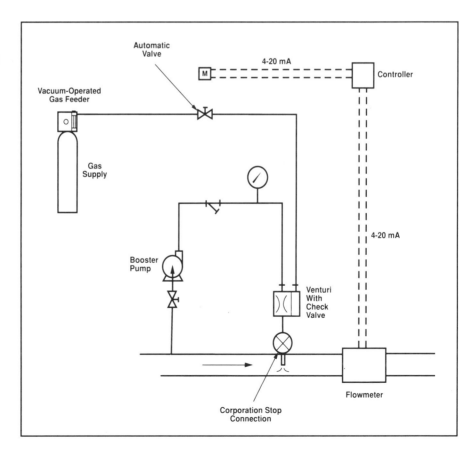

feeder's controller. Flow proportioning control is applicable for both groundwater systems and waste treatment systems that have varying flows and fairly constant demand (Figure 8-5).

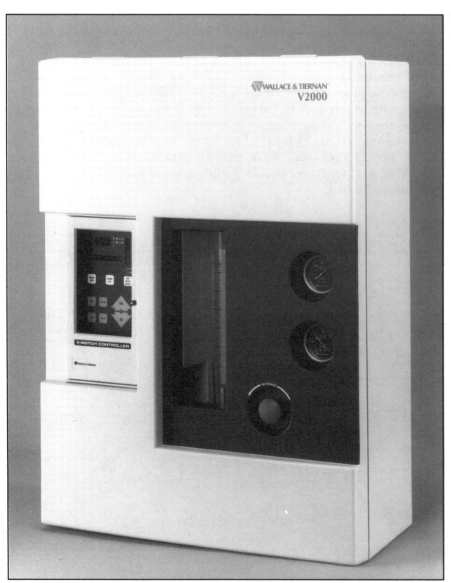

Figure 8-6
Wall-cabinet-mounted automatic gas feeder

Source: Wallace & Tiernan Inc.

Automatic Residual Control Gas Feeder

Where flow proportioning control fails by providing no feedback to the controller, residual control fills the gap (Figure 8-6). Residual control is often called *closed-loop control* since there is feedback to the controller of the result of the change, and, thus, the loop is closed. Automatic residual control requires the use of an instrument external to the gas feeder—a chlorine residual analyzer—to provide a means of continuously measuring the residual chlorine in the water to be treated and sending an electrical signal (4–20 mA) to an automatic controller. The controller will compare the signal (the residual) to the desired residual, or set point, and vary the valve position or valve opening to feed the desired amount of chlorine to reach the desired residual. The difference between flow proportioning control and residual control is that residual control includes feedback of the result of the input. Although this scheme provides the result of the input, it does not recognize the source of the change. Changes in water flow, which will cause a change in residual, typically occur rather rapidly. The residual control scheme is not capable of handling a rapid change of flow and its usefulness under these conditions is limited.

Figure 8-7
Typical residual control system

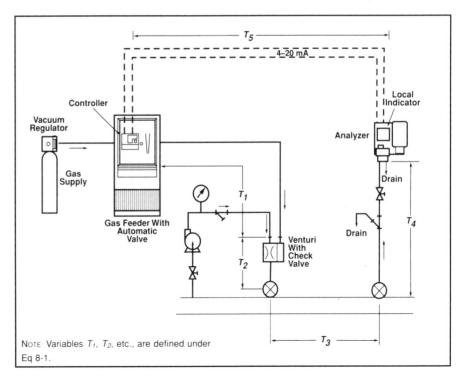

NOTE: Variables T_1, T_2, etc., are defined under Eq 8-1.

Applications for residual control equipment include situations in which the flow does not vary or varies so slowly that the control scheme can recognize the change and respond to make proper corrections to the chlorine feed rate. Water treatment plants with clearwells of sufficient size may be such an application since the demand is fairly constant and the clearwell can provide sufficient holdup to attenuate flow variations. Care must be taken with the use of residual control to ensure proper mixing and treatment time. The sampling point is critical to accurate control since the treatment time after chlorine addition is a factor in the process, but is not measured as an element in the residual control loop (Figure 8-7).

It is important that the time for each step in the closed-loop control process be recognized. Equation 8-1 provides the total time in the control loop as the sum of its components. If time values are determined for each portion, a better understanding of the control scheme can be determined. The variables in Eq 1 are illustrated in Figure 8-7.

$$T_t = T_1 + T_2 + T_3 + T_4 + T_5 \qquad (8\text{-}1)$$

Where:

T_t = the total time in the loop from the time of correction to the time the electrical signal is received at the analyzer

T_1 = the time from valve position change to the time that the gas signal change reaches the venturi in the ejector

T_2 = the time from the mixing of the gas with the water in the venturi until the solution formed reaches the point of application

T_3 = the time from the point of application to the point of sampling

T_4 = the time from the point of sampling to the analyzer

T_5 = the time for the analyzer to process the sample to the time the electrical signal reaches the controller

For practical purposes, the times in the loop that are electrical in nature are insignificant compared to the other factors. The remaining critical times are of importance and their influence on the control result are minimized if

- the control valve is located as close as possible to the venturi
- the venturi is as close as possible to the point of application
- the sample point location is as close to the point of application as the process will permit
- the sample transfer time to the analyzer is minimized by locating the analyzer as close as possible to the sampling point
- the analyzer reads the residual value constantly, not intermittently

Some European utilities, particularly in the United Kingdom, practice triplication by using three sensors (analyzers). This practice is particularly helpful at unmanned sites since three values provide some assurance of proper analyzer performance. The theory of this practice is that if at least two of the analyzers are reading the same, operating parameters are being met. The third analyzer could be checked and calibrated at a convenient time. This way the correct signal is obtained from among the three analyzers and sent to the controller. The use of three analyzers can be adapted for monitoring as well as control conditions.

Compound-Loop Control Gas Feeder

The ideal control scheme can be obtained by taking both the open- or single-loop control and closed-loop control systems and combining them into one scheme, hence the name *compound-loop control* (Figure 8-8). In this control system, two signals, flow and residual, are sent to one controller that combines the two inputs mathematically. The controller compares the resultant to a set point and sends one electrical signal to the automatic valve. The flow signal is the major control input because water flow changes are instantly measured and the flow signal is corrected rapidly. Meanwhile, the residual signal is used as a final trim for the valve. The advantage of this control scheme is that it combines the features of flow proportioning (quick response to changes in flow) and residual control (recognizing residual deviations) and adjusts for both at the same time.

Two sensing instruments, the chlorine residual analyzer and the water flowmeter, provide the inputs to the controller. The controller provides the ability to control to a fixed residual in a variable flow and variable demand condition. The location of the sampling point is critical to optimum operation. This sampling point need not be after the "mandated" contact time but located within a reasonable time after chemical addition. The decrease in residual is a function of the reaction rate and demand. The controller design must provide for lag-time considerations and be able to compensate for variations in flow that will alter contact times for fixed sampling locations. The ideal controller design should have a feature that permits either input to be used independently for valve control in the event of a failure of the flowmeter or residual analyzer.

Compound-loop control may be used in distribution systems where rechlorination is required and at some water treatment plants where flow and residual may vary. Response to rapid flow changes is a hallmark of these systems (Figure 8-9).

Figure 8-8
Floor-cabinet-
mounted
automatic gas
feeder

Source: Wallace & Tiernan Inc.

Ammonia Feed and Operating Situations

Ammonia feed systems generally use only one control scheme. In the chloramination process, the control of the chlorine–ammonia ratio at 3:1 to 4:1 is of paramount importance, the ammonia feeder is tied directly to the chlorine residual and the water flow. Thus, the compound-loop control would be the most logical choice for the feeder. Some facilities, however, may use flow proportioning control and manually set the chlorine–ammonia ratio. An example of this type of application is a well-water chlorination system where ammonia is added to well water immediately after chlorination. For this application, manual ammoniation control with a semiautomatic on/off system is appropriate.

The nature of ammonia feed systems is such that a more problem-free performance may be attainable by using a pressure feed system. Although using pressure feed for ammonia sacrifices the safety of the all-vacuum system, the relative safety of handling ammonia, compared to chlorine, combined with the service-free system offered by ammonia pressure feed, more than balances the loss of all vacuum safety of feeding ammonia.

Therefore, ammonia feed systems will use automatic control schemes and may use either manual or semiautomatic systems. Ammonia monitors may also be used to provide a signal with flow input for compound-loop control. All systems require the maintenance of the proper chlorine–ammonia ratio (Figure 8-10).

Figure 8-9
Compound-loop
control

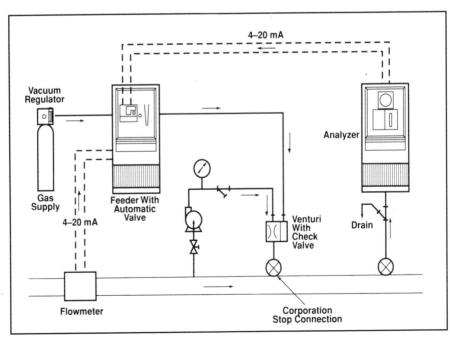

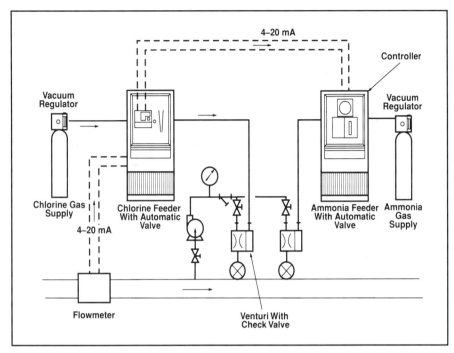

Figure 8-10
Chlorine/
ammonia feed
systems

Chemical Feed Pump Control

Chemical feed pumps (Figure 8-11) feeding either hypochlorite or ammonia solutions may be automated the same control schemes used in gas feeders. Thus, control can be manual, semiautomatic, flow proportioning, residual, or compound loop. The pump can be started and stopped manually or semiautomatically in the same fashion as the gas feeders previously described. The feed rate is set manually to attain the proper dosage by adjusting the stroke length, pulse, or pump speed, depending on the type available and following the manufacturer's instructions.

For automatic control with either flow proportioning, residual, or compound-loop control, most modern chemical feed pumps can receive a 4- to 20-mA signal from the appropriate controller. The same controller that is used for gas feeders can be used to provide the correct input signal to the chemical feed pump. Most commercially available chemical feed pumps use variable DC drives that receive a 4- to 20-mA signal from the controller. The pump feed rate varies in response to the signal automatically. Most commercially available pumps adjust the feed rate by varying the pump speed, although, depending on the manufacturer, there may be other methods. It is always critical to have the pump properly calibrated following the manufacturer's instructions and to

Figure 8-11
Chemical feed
pumps

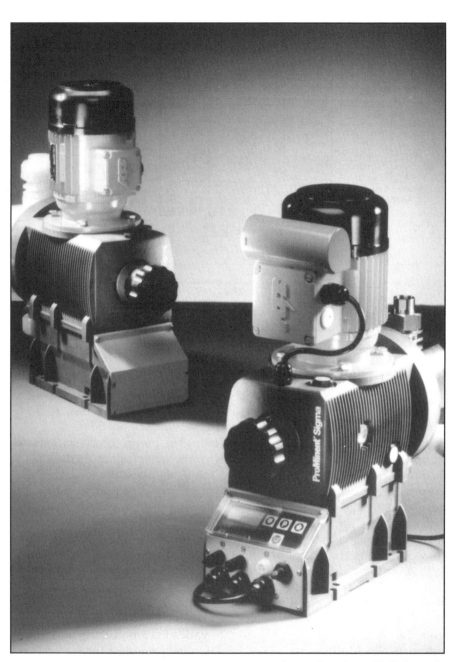

Source: Prominent Fluid Controls.

know the strength of the solutions being fed so that the feed rate or dosage can be properly established.

Just as it is important to have the pump calibrated, it is also important to check the control elements on a regular basis and provide calibration at scheduled intervals. Such maintenance applies to the flowmeter, analyzer, pH meter, and other devices used in the disinfection control scheme. If standby equipment is on-site, then regularly scheduled use of this equipment should be established and followed.

Glossary

ABS Acrylonitrile butadiene styrene. A high-impact plastic used in the manufacture of chlorinators and ammoniators.

ACGIH American Conference of Governmental Industrial Hygienists.

advanced oxidation processes Processes that use combinations of oxidation techniques, such as ozone and hydrogen peroxide.

alkalinity Neutralization capability of water, expressed in mg/L.

ammonia A chemical compound used in the treatment of water to form chloramines.

APHA American Public Health Association.

aqueous ammonia A water solution of ammonia, commercially about 30 percent strength.

ASME American Society of Mechanical Engineers.

atm Atmosphere.

AWWA American Water Works Association.

BOCA Building Officials and Code Administrators of America.

Breakpoint The point in the chlorination process where the combined chlorines are oxidized by free chlorine.

Buffer A chemical compound added to water that maintains the acidity or alkalinity of a solution at a certain level.

$C \times T$ The combination of chlorine residual and exposure time that is used to establish the effectiveness of the disinfection process.

CFR Code of Federal Regulations.

CGA Compressed Gas Association.

chelating agent A chemical compound that acts to bind with certain compounds in water and prevent interaction with other chemicals.

CHLOR REP A North American emergency response organization operated under the auspices of the Chemical Manufacturers Association.

chloramination The process in water treatment in which a chlorine/ammonia compound is used as the disinfecting agent.

chloramines The product of the reaction between chlorine and ammonia compounds.

chlorate A decomposition product of hypochlorite and of the chlorine/chlorite reaction that produces chlorine dioxide.

chlorination The process in water treatment of adding chlorine (gas or hypochlorite) for purposes of disinfection.

chlorine The element that produces hypochlorous acid on reaction with water.

chlorine dioxide An oxidant produced by the reaction of chlorite and chlorine.

chlorite A precursor for the chlorine dioxide reaction.

cholera A debilitating illness caused by waterborne bacteria.

compound-loop control Control of chlorine feed by both flow and residual control.

CPVC Chlorinated polyvinyl chloride.

Cryptosporidium An organism found in water systems that causes intestinal distress.

CT Contact time.

cylinder A metallic container used to contain compressed chlorine or ammonia gases.

DBP *See* disinfection by-product.

dechlorination The removal of chlorine by any one of several reducing agents.

demand The requirement for chlorine as determined by the dissolved organics, inorganics, and bacteria and other organisms.

dichloramine A product of the chlorine and ammonia reaction.

differential-pressure regulator A device used to maintain a fixed pressure in a fluid line.

dip tube An extension of a container's valve inlet to some remote point; usually used for liquid rather than gas access.

disinfection The act of removing pathogenic bacteria.

disinfection by-products The reaction products between chlorine and precursors, including THMs, haloacetic acids, aldehydes, etc.

dosage The amount of additive to water expressed as mg/L or ppm.

DOT US Department of Transportation.

DPD N,N-diethyl-p-phenylenediamine. A chemical reagent used in chlorine residual testing.

drop dilution method A method used to measure high levels of chlorine residuals.

dysentery A disease caused by waterborne bacteria.

elemental chlorine A term used to define compressed chlorine gas.

emergency kits Kits designed by The Chlorine Institute and used to contain chlorine leaks in cylinders, ton containers, tank cars, and tank trucks.

expansion chamber The vessel used with liquid chlorine lines to protect against hydrostatic rupture.

FAC *See* free available chlorine.

FACTS Free available chlorine test, syringaldazine.

fire codes Codes produced by the three code organizations that govern the installation of hazardous materials.

flexible connector Copper tubing used on pressure lines from chlorine storage vessels.

free available chlorine The sum of hypochlorous acid and hypochlorite ions expressed in terms of mg/L or ppm.

FRP Fiberglass reinforced plastic.

GAC Granular activated carbon.

gas feeder Device used to feed chlorine and ammonia gases to water.

Giardia lamblia An intestinal organism from mammals that appears in improperly treated drinking water.

groundwater Water available from below the surface of the ground and obtained by using wells.

hard water Water containing large amounts of calcium and magnesium salts.

HazMat Acronym for hazardous materials, used to define personnel engaged in containing and/or safely handling such materials.

hydrolysis The reaction of chlorine with water.

hypochlorite ion The cation component of sodium hypochlorite.

hypochlorous acid The reaction product of hydrolysis.

IDLH Immediately dangerous to life or health.

IFCI International Fire Codes Institute.

intermediate chlorination Chlorination at a point during the water treatment process.

iodometric test method A residual chlorine analysis technique used primarily in the laboratory.

jar test Laboratory test on water samples used to help predict the results of a process.

javelle water A name for sodium hypochlorite solutions; named after a town in France where it was first used.

liquid bleach Sodium hypochlorite solutions.

liquid chlorine Compressed pure elemental chlorine held in pressurized containers.

liquid feeder A mechanical device (usually a positive displacement pump) used to feed liquid chemicals.

log reduction The reduction by a power of 10 of an amount of material (e.g., bacteria and viruses).

manual control The control of chemical feed by manual means.

material safety data sheet Chemical data sheet usually provided with the shipment by manufacturers and shippers of chemicals.

mg/L Measurement of the concentration of chemical dissolved in water; milligrams per litre. *See* ppm.

µg/L Micrograms per litre; measurement of the concentration of a material dissolved in water. *See* ppb.

min Minute.

monochloramine The result of the reaction between chlorine and ammonia that contains one chlorine atom.

MSDS *See* material safety data sheet.

National Code Fire and building codes produced by BOCA.

NFPA National Fire Prevention Association.

NIOSH National Institute for Occupational Safety and Health.

nurse tank An ammonia container used primarily in agricultural applications.

O&M Operations and maintenance.

once-through system In cooling water systems, a process flow of water wherein the water is returned to its source after it is used for heat removal.

OSHA Occupational Safety and Health Administration.

oxidizing agent A chemical agent that gains electrons in a chemical reaction.

ozone A triatomic form of oxygen.

PAC Powered activated carbon.

PAO Phenylarsine oxide. A titrant used in determining chlorine residuals.

PE Polyethylene.

PEL Permissible exposure limit.

pH A measure of the hydrogen ion concentration in water.

pigtail The flexible copper tubing connector used for hookup to cylinders and containers.

polyphosphates Chemicals used as chelating agents in water treatment.

postchlorination Chlorination practiced after the other treatment processes, usually in the clearwell prior to pumping to the distribution system.

powder bleach Calcium hypochlorite.

PP Polypropylene.

ppb Parts per billion, numerically the same value as $\mu g/L$.

ppm Parts per million, numerically the same value as mg/L.

proportioning control Control of one flow stream in a fixed proportion to another flow stream, such as flow proportioning in water treatment.

PRV Pressure-reducing valve.

PTFE Polytetrafluoroethylene; generic term for Teflon®.

PVC Polyvinyl chloride.

rechlorination A chlorination practice, usually in the distribution system, to raise the residual level. Sometimes used with purchased water.

recirculated water In a cooling water system, water that is constantly recirculated as a coolant, e.g., in cooling towers.

reducing agent In chemical reactions, that which gives up electrons, e.g., sulfur dioxide.

reliquefaction The recondensation of a gas after it has been liquefied.

residual The measurement of a chemical in water (e.g., chlorine) after treatment.

residual control Automatic control of chlorine feed to maintain a constant chlorine residual.

RQ Reportable quantity, minimum quantity established for releases of materials above which reports to the appropriate authorities must be made.

Safe Drinking Water Act Act passed by Congress to protect the quality of the drinking water in the United States.

SARA Title III Section 3 of the Superfund Amendments and Reauthorization Act of 1986, which deals with the emergency planning, emergency release notification, community right to know, and toxic chemical release reporting.

SCBA Self-contained breathing apparatus.

SCR Silicon-controlled rectifier; a variable DC drive.

scrubbers Mechanical equipment used to remove undesirables in a gas discharge.

SDWA *See* Safe Drinking Water Act.

semiautomatic control Control of chemical feed that operates with no feedback and with intermittent signal inputs.

shock treatment The periodic addition of chlorine to treat cooling water for fixed durations.

soft water Water considered to have no calcium or magnesium salts; no hardness.

sonic regulator A regulator used in a gas vacuum that has the flow of gas at the speed of sound through the control orifice.

Standard Code Fire and building code produced by the Southern Building Code.

standard methods The name applied to the analytical methods used in water treatment and published by AWWA, APHA, and WEF in *Standard Methods for the Examination of Water and Wastewater.*

STEL Short-term exposure limits.

stoichiometry The correct amount of chemicals used in a reaction as determined by the chemical equation.

strong acid An acid that is almost completely dissociated, usually a mineral, inorganic acid.

substitution reaction A reaction in which there is no change of valence in the reactants.

superchlorination Chlorination in water treatment with an exceptionally high dosage, usually applied to achieve breakpoint.

surface water A water source that is above ground, e.g., lakes and rivers.

Surface Water Treatment Rule Regulations issued by the USEPA to meet the requirements established by the SDWA.

SWTR *See* Surface Water Treatment Rule.

tantalum Metallic element resistant to chlorine used in chlorinator manufacture.

TCE Trichlorethylene.

Ten State Standards Standards established for the design and operation of water treatment plants.

THM *See* trihalomethane.

titanium Metallic element that will react with dry chlorine. If used for chlorine service, must be in the presence of water or water vapor.

TLV Threshold limit value.

TOC Total organic carbon.

ton container Steel vessel used to hold 2,000 lb (906 kg) of compressed chlorine.

TPQ Threshold planning quantity. That amount above which appropriate emergency plans must be made.

trichloramine Compound of chlorine and ammonia, sometimes referred to as nitrogen trichloride.

trichloroethane One of the trihalomethanes.

trihalomethane A group of halogen compounds with one carbon atom and various combinations of chlorine and bromine.

TWA Time-weighted average. Used to evaluate over a given time period the allowable exposure to various materials considered hazardous to health.

typhoid Disease caused by a waterborne bacterium.

Uniform Code A model building and fire code used to establish regulations for the installation of chemical feed systems. Produced by the Western Fire Chiefs Institute.

US Environmental Protection Agency The US agency charged with the responsibility of establishing the criteria for and the policing of discharges to the environment.

USEPA *See* US Environmental Protection Agency.

vacuum regulator Mechanical device used to drop the gas pressure of chlorine or ammonia to a vacuum level to improve safety in handling these materials.

valence The positive or negative value of an element that reflects its oxidizing potential.

vaporizer A piece of equipment used to convert chlorine and ammonia liquids to a gas.

veliger Name applied to immature zebra mussel.

venturi The reduction in pipe diameter used to create a vacuum for operation of a gas chlorinator or ammoniator.

water softener Mechanical device containing ion exchange resins and used to remove undesirable calcium and magnesium ions from water.

weak acid An acid (acetic, phosphoric) that is incompletely dissociated in water solution.

WEF Water Environment Federation. An association dedicated to providing a clean effluent from a wastewater treatment plant.

WPCF Water Pollution Control Federation. Former name of the WEF.

zebra mussel Crustacean, native to European waters, that has multiplied rapidly after an accidental release in North American waters and is causing difficulty in water intakes.

Typical Calculations

Some common calculations used by water treatment plant designers and operating personnel to determine solutions to typical disinfection system problems are presented in this appendix. Calculations are performed both in metric and US units.

1. What dosage of chlorine will be required to treat water with a demand of 0.6 mg/L (ppm) to leave a residual of 0.5 mg/L (ppm)?

   Equation: Dosage = Demand + Residual

   Metric Units: Dosage = 0.6 mg/L + 0.5 mg/L = 1.1 mg/L

   US Units: Dosage = 0.6 ppm + 0.5 ppm = 1.1 ppm.

2. If the water flow is 315.4 m^3/h (2.0 mgd) and the dosage is to be 1.1 mg/L (1.1 ppm), what feed rate of chlorine gas will be required?

   Equation: Water Flow × Dosage × Unit Conversion Factor = Feed Rate

   Metric Units: Feed Rate = 315.4 m^3/h × 1.1 mg/L × 1,000 L/m^3 × 1 g/1,000 mg = 347 g/h or 0.347 kg/h

   US Units: Feed Rate = 2 × 10^6 gal/d × 8.345 lb/gal × 1.1 parts Cl_2/10^6 parts H_2O = 18.3 lb/d

NOTE: Chlorinators and ammoniators are rated in either grams or kilograms per hour or pounds per 24-h day.

3. If the water flow rate is 315.4 m³/h (2 mgd) and the water contains 0.15 mg/L (ppm) of iron, 0.05 mg/L (ppm) of manganese, and 0.12 mg/L (ppm) of sulfide, what would the total dosage of chlorine be to oxidize the iron and manganese and convert the sulfide to sulfate?

Contaminant	Amount, mg/L	Cl_2 Required per Part of Contaminant	Cl_2 Required
Fe^{+2}	0.15	0.6	0.09
Mn^{+2}	0.05	1.3	0.065
S^-	0.12	6.2	0.744
Total Dosage			0.899

4. What would the feed rate of chlorine be to oxidize the contaminants in question 3?

Equation: Water Flow $\times$ Dosage $\times$ Unit Conversion $=$ Feed Rate

Metric Units: Feed Rate $=$ 315.4 m³/h $\times$ 0.899 mg/L $\times$ 1,000 L/m³ $\times$ 1 g/1,000 mg $=$ 223.5 g/h or 0.2835 kg/h

US Units: Feed Rate $=$ 2 $\times$ 10⁶ gal/d $\times$ 8.345 lb/gal $\times$ 0.899 parts Cl^2/10⁶ parts H_2O $=$ 15.0 lb/d

5. If the water flow is 315.4 m³/h (2 mgd), has zero demand, and contains no chlorine or ammonia, how much of each chemical must be added to provide a 0.5 mg/L (ppm) residual of monochloramine?

Metric Units: 315.4 m³/h $\times$ 0.5 mg/L $\times$ 1,000 L/m³/h $\times$ 1 g/1,000 mg $=$ 157.7 g/h monochloramine

Following the stoichiometric quantities for monochloramine from Eq 3-5 on page 28, one part of monochloramine requires 1.39 parts of chlorine and 0.333 parts of ammonia.

Therefore, 219.5 g/h of Cl_2 and 52.6 parts of NH_3 are required.

US Units: 2 $\times$ 10⁶ gal/d $\times$ 0.5 parts Cl_2/10⁶ parts H_2O $\times$ 8.345 lb Cl_2 $=$ 8.345 lb/d monochloramine

Following the stoichiometric quantities for monochloramine 8.345 lb/d monochloramine $=$ 11.62 lb Cl_2 and 2.78 lb/d NH_3

NOTE: As shown in Eq 3-5, not all the chlorine added is used in the monochloramine product, thus more chlorine must be added than required by the monochloramine.

6. A chlorine feed installation feeds 12 lb of chlorine gas per day (18.9 g/h). How much of each sodium hypochlorite solution is required to be fed if the solution strengths are 13 percent, 10 percent, and 5 percent, respectively, to provide equal oxidizing capability?

From Table 2-3 (page 16)

Metric Units:
13 percent: 1 L of solution/129.5 g Cl_2 × 18.9 g/h
  = 0.149 L/h

10 percent: 1 L of solution/99.6 g Cl_2 × 18.9 g/h = 0.19 L/h

5 percent: 1 L of solution/50.4 g Cl_2 × 18.9 g/h = 0.375 L/h

US Units:
13 percent: 1 gal solution/1 lb Cl_2 × 12 lb/d Cl_2 ×
  1 d/24 h = 0.5 gal/h

10 percent: 1 gal solution/0.83 lb Cl_2 × 12 lb/d Cl_2 × 1 d/24 h
  = 0.6 gal/h

5 percent: 1 gal solution/0.42 lb Cl_2 × 12 lb/d Cl_2 × 1 d/24 h =
  1.19 gal/h

7. The calculation in question 2 dealt with chlorine gas feed. If a solution of 13 percent sodium hypochlorite was being fed instead of chlorine gas, what feed rate would be required?

Equation: Water Flow × Dosage × Conversion × Equivalent Solution Strength = Hypochlorite Feed

Metric Units: 315.4 m^3/h × 1,000 L/m^3 × 1.1 mg/L ×
  1 g/1,000 mg × 1 L/129.5 g = 2.68 L/h

US Units: $2 × 10^6$ gal/d × 1 d/24 h × 8.345 lb/gal × 1 gal solution/1.08 lb Cl_2 × 1.1 lb Cl_2/10^6 lb water = 0.71 gal solution/h

8. If a 5 percent hypochlorite solution were used in the calculation in question 7, what would the hypochlorite feed rate be?

Metric Units: 2.68 L/h × (129.5 g/L)/(50.4 g/L) = 6.9 L/h

US Units: 0.71 gal/h × (1.08 lb/gal)/(0.42 lb/gal) = 1.82 gal/h

9. If aqueous ammonia solutions were used instead of ammonia gas for the calculation in question 4, what would be the feed ratio of 30 percent aqueous ammonia and 13 percent sodium hypochlorite solutions?

Metric Units: The calculation from question 4 showed that a feed note of 157.7 g/h of monochloramine was needed from 219.5 g/Cl$_2$ and 39.9 g NH$_3$.

Therefore, for 219.5 g Cl$_2$:

(219.5 g Cl$_2$/h) / (129.5 g Cl$_2$/L) = 1.69 L/h
and for 39.9 g NH$_3$: (39.9 g NH$_3$/h) / (30 g NH$_3$ / 100 g solution) × (1 g solution/0.8957) = 119 L/h

US Units: The calculations from question 4 showed that a feed of 8.345 lb/d of monochloramine was required from 11.62 lb/d of Cl$_2$ and 2.78 lb/d of NH$_3$.

For 11.62 lb/d of Cl$_2$: 11.62 lb Cl$_2$/d × 1 d/24 h × 1 gal solution/1.08 lb Cl$_2$ = 0.45 gal/h

For 2.78 lb/d of NH$_3$: 2.78 lb NH$_3$/d × 1 d/24 h × 100 lb solution/30 lb NH$_3$ × 1 gal/7.48 lb = 0.05 gal/h

Index

NOTE: An *f.* following a page number refers to a figure; a *t.* refers to a table.

Breakpoint reaction, 5, 33–36
  curve, 34, 35*f.*
Bromine, 50
Building Officials and Code Administrators of America, 92

C *x T* concept, 39, 45–47
  values for inactivation of virus, 46*t.*
  values for reduction of *Giardia*, 46*t.*
Calcium hypochlorite, 3, 18–19, 27
  oxidizing capability, 28, 28*t.*
Calculations, 157–160
Canada, 6
CGA, 88
Chemical feed pumps, 114–115, 145–147, 146*f.*
  hydraulic drive, 115, 116*f.*
  maintenance and service, 117
  mechanical drive, 114–115, 115*f.*
  operation, 115–117
  pressure, 115–117
  pump calibration, 115, 116*f.*
Chemical Manufacturers Association, 88, 97
CHEMTREC, 97
Chloramination, 4–5, 44. *See also* Trihalomethanes
  as part of water treatment process, 2
  breakpoint reaction, 5, 33–36
  future, 7–8
Chloramines, 28–29, 43
  mono- and dichloramine distribution, 30, 31*f.*
  odor thresholds, 29, 29*t.*
  sample dosage calculations, 158
  taste thresholds, 29, 30*t.*
Chlorate, 17–18, 18*f.*, 19*f.*
Chloride of lime, 3, 27
Chlorinated polyvinyl chloride, 10, 105
Chlorination, 1. *See also* Trihalomethanes
  application points, 47–51
  and control of waterborne diseases, 3, 6
  dechlorination, 52–53, 53*t.*
  early use, 3–4
  engineering considerations, 57–59
  future, 7–8
  groundwater, 47, 54–55
  intermediate, 50
  nondisinfection applications, 3

163

Hypochlorites, 3, 4, 24
  and hypochlorous acid, 25, 26f.
  sample dosage calculations, 159
Hypochlorous acid, 3, 15, 24, 25, 27
  and hypochlorite, 25, 26f.
  as weak acid, 25

IFCI. *See* International Fire Codes Institute
Induction mixers, 104, 106f.
Intermediate chlorination, 50
International Fire Codes Institute, 92
Iodometric analysis method, 63–64
Iodometric electrode analysis method, 63–64

Javelle water. *See* Sodium hypochlorite

Kel-F, 105, 111
Kynar, 10

Latin America, 6
Lincoln, England, 4
Liquid bleach. *See* Sodium hypochlorite
Liquid chlorine, 4
Lithium hypochlorite, 18
Louisville, Kentucky, 4

MAC. *See* Maximum allowable concentrations
Maidstone, England, 4
Manual control gas feeders, 134, 135f., 136f.
Material safety data sheets, 20, 97–98
Maximum allowable concentrations, 123
Membrane filtration, 7, 8
Microbiological safety, 8
Middelkerke, Belgium, 4

National Code, 92, 125
National Fire Protection Association, 92
National Institute of Occupational Safety and Health, 11
Netherlands, 5–6, 7
NFPA. *See* National Fire Protection Association
Niagara Falls, New York, 4
NIOSH. *See* National Institute of Occupational Safety and Health
Nitrates, 32
Nitrites, 32
North America, 6